TRADE UNION MOVEMENT IN BANGLADESH

TRADE UNION MOVEMENT IN BANGLADESH

DR. MD. GHOLAM KIBIRA
Edward College, Khulna, Bangladesh

DR. ANIL BHUIMALI
Department of Economics
University of North Bengal
West Bengal, India

DR. KANAK KANTI BAGCHI
Department of Economics
University of North Bengal
West Bengal, India

SERIALS PUBLICATIONS
NEW DELHI

First Published - 2006

ISBN : 81-8387-013-9

Published by

SERIALS PUBLICATIONS

4830/24, Prahlad Street, Ansari Road
Darya Ganj, New Delhi-110002 (India)
Phone : 23245225. Fax : 91-11-23272135
E-mail: serials@satyam.net.in

Dedicated to Our Parents

CONTENTS

PREFACE

Trade Unions were formed first by the Factory Workers of England against the tyranny and neglect of the factory owners. The objectives of the trade unions, among others, are to fix wages and to deal with the employers, to regulate the wage rate according to profit earned by the factory owners. The present book an extended version of a research investigation (a Ph.D. Thesis), is timely and relevant as trade unions emerge to become stronger forces that industrial managers cannot undermine. The unions significantly influence working conditions, management and leadership and the socio-economic status of workers. At the macro level, these unions have impact on the rate and the nature of economic growth and development.

The plight of trade unions and the industrial workers is a major concern for policy makers and development planners to address. The in-depth analysis that was made on the trade union movement particularly in the jute industry, will lead to policies and intervention by the government, the private sector, the civil society and other interest groups which have stakes in the industry. Entry variables for community based intervention that is dovetailed to fit the unique Bangladeshi experience can be designed using the results of the study.

The core of the study has been the representation of two case studies, one with nationalized jute mills and the other with private jute mills of Bangladesh. Data gathered from nationalized jute mills show that average rate of growth of profit of the public sector jute mills, as a whole, stands at 12.5 per cent. But the reality is that the mills are incurring huge losses over the period. Insecurity of job and low wages have made the workers less efficient and hence less productive. Naturally the jute mills are let in a position to secure positive level of net profit. The picture in private jute mills is more or less the same in terms of production, productivity, export, accumulation of profit or loss incurred. But the socio-economic conditions of workers of private jute mills are more worse than that of nationalized jute mill workers. Not only of that workers of private sector jute mills are less powerful and their collective bargaining power in raising wages and other benefits has been worsening continually.

Md. Gholam Kibria
Anil Bhuimali
Kanak Kanti Bagchi
Department of Economics
University of North Bengal

Chapter - 1
THE PROBLEM

Introduction

Before tracing out the evolution of trade union movement in the Indian subcontinent during the British rule let us begin with the definition and meaning of 'Trade Union'.

Sidney and Beatrice Webb defined trade union as "a continuous association of wage earners for the purpose of maintaining or improving the conditions of their working lives" (Webb & Webb, 1894: 1). The World Book of Encyclopedia has given a comprehensive but meaningful definition of trade union. "Trade unions are organizations of workers that seek to improve their members pay and working condition (World Book of Encyclopedia, 1993: 328). V.V. Giri has given a worth mentioning definition of trade union. He defines that "trade unions are voluntary organizations of workers formed to promote and protect their interests by collective action" (Giri, 1962:1). According to G.D.H. Cole, "a trade union is an association of workers in one or more occupation-an association carried out mainly for the purpose of protecting and advancing the members' economic interests in connection with their daily work" (Cole, 1962: 13). Edwin B. Filippo defined trade union as "a labour union or trade union is an organization of workers formed to promote, protect and improve, through collective action, the social, economic and political interests of its members (Filippo, 1980: 378). The Industrial Relations Ordinance, 1969, which is still in operation (with modifications) defines trade union as follows: "Trade Union means any combination of workmen or employers formed primarily for the purpose of regulating the relations between workmen and employers or workmen and workmen or employers and employers, or for imposing restrictive coordination on the conduct of any trade or business and includes a federation of two or more trade unions". It can be said that trade union is a continuous association of workers or employees formed to promote and protect their vocational interest. They had to be united to form an association from the earliest period of the growth of labour class as they were suppressed and oppressed by the employer, ruthless administrations and the Government. Government had always support for the employer at that time. This was the trend all over the world.

So the workers were compelled to think to be united for doing something for common interest. As factory industry had been started much earlier and industrial revolution took place in 18th century in Great Britain, first attempt of joint activities were observed there. The name 'trade union' in Great Britain, seems to have been first used to describe joint activity by little trade clubs in defiance of the common interests of the workers in a whole town or area. Trade clubs were regarded as primarily a friendly society for mutual aid through the provision of friendly benefits in sickness or accident (Cole, 1962: 18).

Though factory Industry emerged rather late in Bengal and India as well during the second half of 19th century, it had its heritage of cottage industry long back. The different classes of artisans of that time used to take up their grievances against their employer to the *Panchayat*. Later on these trade clubs and guilds or *Panchayats* were the beginning of workers unification move.

Before 1850s the only industry in the modern sense was coal-mining. Number of development took place between 1850 to 1869 which led to the first stage of 'industrial revolution'. A start was made at building railways around Bombay and

Calcutta, and more importantly in 1954 Calcutta was linked by rail to the coal-mining centers, Raniganj. In the same year a Scottish entrepreneur established the first jute mill in Bengal and before that a cotton textile mill had been established in Bombay in 1951 (Crouch, 1979:13). During the next 60 years industries developed through a gradual process. In 1914 there were 264 cotton mills working in India employing some 260,000 operatives. Jute industries had been developed exclusively in Bengal. There were more than 60 mills employing about 2,00,000 workers. By 1914 about 6,00,000 people were employed on the railways (Karnik, 1960: 7-8). The development of industries resulted in a continuous increase in the size of the industrial workforce. But their misery was beyond description.

A factory worker M.N. Lokhande was able to arrange a meeting of labourers of Bombay on 26 April, 1884 where ten thousand workers attended the meeting with legitimate demand including a weekly rest day. In response to the workers demand mill owners of Bombay granted a weekly holiday. It was the first mass meeting of workers in Bombay and India as well. Inspired with this success, Lokhande formed a workers' organization called the Bombay Mill Hands Association. Lokhande actually laid the foundation stone for the trade union movement in the subcontinent', although it was not a trade union organization in the truest sense (Ahmed, 1935:13). There was no working hour, no holiday, no beginning and no closing time. Therefore, it was too difficult to be united to form any association or union. Employer had the paternalistic attitude at that time and used to control everything in iron hand. Most employers had seventeenth and eighteenth centuries' idea about their relationship with the employers. The attempt to form a union was regarded as an act of treachery, disloyalty and ingratitude (Karnik, 1960:307). With a very few exceptions the formation of trade unions has been looked upon by management with suspicion and hostility, and that sometimes it has done its best to crush the newly formed unions. This was the case during early years of industrialization in the United Kingdom, and the newly industrialised countries may not be an exception to this (Rahman, 1969: 159).

According to Marx, the British working class were the first-born sons of modern industry. So they were naturally the pioneers of trade unionism. Trade union organization is the outcome of the economic struggle of the class of working men against the new class of capitalist employers, against the competition of all which reigns in modern civic society (Quoted in Allen Hutt, 1962: 7).

In the early years of factory system in U.K. social significance of the formation of trade unions were the conduct of strikes against low wages, long hours and abominable conditions (Hutt, 1962: 7). By the end of eighteenth century trade unions had begun to take root, in the shape of local trade clubs which usually met in public houses. For the most part, these clubs had developed aristocracy among the artisans. However, from 1792 trade union activity began to spread among the key section of the new factory workers, the Lancashire cotton spinners and this was a potential threat to the rising capitalist employers. A wider unity, a more universal solidarity began to support the parochial version of the local trade club. So there was series of suppression, oppression and torture against the formation of trade union. During quarter of a century this reign of anti-union terror lasted and the trade union was really born.

Specially when the coal fields declared strikes, there were civil wars in miniature. For example, in 1831-32, Darham was in turmoil, and marines and cavalry were drafted in to break the strikes and the union. Troops were also called. The Glasgow Cotton-spinners, whose union had been formed secretly.

Industrial unions were a later growth of the last two decades of the nineteenth century, with the growth of mass production industries and of the railways and docks and mines, it became necessary to think of a different form of organization in which unskilled and ordinary workers could join without any difficulty. Industrial unions were found to be more effective in conflicting with employers. Craft unions did not disappear, but in new mass production industries they had to yield place to industrial unions. Another type of union which grew in the later period was the general union catering to the needs of workers employed in dispersed industries or by a large number of employers (Karnik, 1982: 36-37).

In United States of America workers organisation existed even before their independence. They were mostly benevolent societies for mutual help than modern trade unions. Modern type of trade unions began to be organized in the last decade of the eighteenth century. They were organized on a craft basis. Employers opposed them as illegal conspirators. Early in the nineteenth century they succeeded in getting some court decision against unions and more particularly against strikes (Karnik, 1982: 43). So the early stage of trade union movement of all the countries are the same. They had to fight for working hours, fair wage and right to organisation.

There were several attempts of the workers to get united to form their association, but due to obstinate resistance from the employers the organisations could not get a solid stand. Philadelphia shoemakers made the first attempt in 1792, the organisation lasted less that one year, it was organised in 1794 and was still in existence in 1806. New York Printers organised a permanent Typographical Society in 1794. Between that date and 1818 shoemakers established fairly permanent organisations in different important cities (Rayback, 1966: 54-55). National Labour Union (NLU), the first permanent organisation of labour in America on a national seal was formed in 1866 (ibid. : 116). Later on a new organisation, 'Knights of Labour' was formed in December 1869 (ibid.: 143). The Knights more concretely stood for an eight-hour day, equal pay for equal work, public ownership of utilities and establishment of cooperatives. By 1886 they had a membership of over eight lacs. Another new organisation which took place of the Knights was the American Federation of Labour founded in 1886. It concentrated its attention on securing higher wages and better conditions of work through collective bargaining with employers. This is called, pure and simple trade unionism. It came to be known as bread and butter unionism, that is a trade unionism which exerted itself to secure the immediate demands of workers in relation to their jobs such as wages, other benefits, hours of work (Karnik, 1982: 44).

The labour programme of Franklin D. Roosevelt's first administration has been recognised in itself as the most important effort to advance labour welfare in the nation's history. It provided labour with relief from economic distress, assured it some security during periods of unemployment and in old age, and gave it an opportunity to help itself by guaranteeing its right to organize and bargain collectively. The programme helped to create a new organisation – The Congress for Industrial Organisation (CIO) (Rayback, 1966: 346). Moreover, with the industrial development a number of mass production industries came into being in the twenties and thirties. They employed a large number of unskilled workers who could not be fitted into the old craft unions. It was the crying need of the time to organise industrial unions. Unskilled workers did not find a proper place for themselves in the craft union-dominated AFL. So in 1938 CIO was formed. The split lasted for seventeen years. The two organisations came together and in order to satisfy both the parties it was decided to name the reunited organisation AFL-CIO (Karnik, 1982: 46).

The scenario of trade union movement all over the world especially in Asian and in African countries was almost the same. The first labour union was organized in Bangkok in 1877 – only fourteen years after the birth of first Japanese Union and which was incidentally the first Asian Trade Union. But due to immaturity, financial crisis and arrogant attitude of both employer and Government, trade union movements could not develop in the real sense of the term. It was only in 1932 that the tramway workers of Thailand launched a strike when the employees refused to recognise the union and Government also refused to intervene (Ahmad, 1978: xx). Trade union activity was started after 1870 and they had also to face obstinate resistance from both the employer and the Government. Trade union in the Philippines who first organised in 1902 when a union of printers and lithographers was established. Cigar workers' union was formed in 1908 but none of those unions survived because of the hostile attitude of the government (ibid.:xx). The first trade union in Indonesia was formed in 1908 by the railway and Tramway employees. Trade union activities began in Singapore and Malaysia in 1916. Between 1836 and 1914 Malaysia saw a wave of strikes which mostly meant for increasing wage. Majority of strikes took place in Singapore (ibid.: xxi). Strike is one of the media to show the resentment of labourers against any authority to show their grievances. Sidney & Beatrice Webb provided a notable information regarding strike, that first took place in 1490 B.C. (Sidney & Beatrice Webb, 1894: 2).

> ...For instance, find in the revolt, 1490 B.C., of the Hebrew brick makers in Egypt against being required to make bricks without straw, a curious precedent for the strike of the Stalybridge Cotton Spinners, A.D. 1892, against the supply of bad material for their work.

Statement of the Problem

Before 1947, not such remarkable industrial development took place in this part of Bengal. Trade unions are essentially the produce of modern large-scale industry (Karnik, 1960: 2). After partition some industrial units had been established of which ownership of Bengalees was minimum (Farouk-1983: 84-85). The then East Pakistan had different problems also in the way of quick industrialisation. She had shortage of capital and technical know how and efficient entrepreneurs. Despite all the limitations, by 1967-68 total numbers of factories were more than 3,500 (Farouk, 1983: 84).

Immediately after the liberation of Bangladesh in 1971, non-Bengali efficient entrepreneurs left the country. Bangladeshi entrepreneurs got the full opportunity but they were lacking industrial and commercial knowledge. They were not known to tricks and tactics of business. Moreover their attitude was unfavorable as they used to maintain paternalistic attitude with the employees because they lacked industrial traditions. Not only the management of the private establishments but also that of the public undertakings hold uncongenial outlook towards the union. But in recent times attitude of the employer has changed a lot. In some cases labourers and the trade union leaders are doing some acts detrimental to the objective of the very production unit where they are working. During our survey in the mills management personnel of both public and private sector enterprises have provided some information regarding the behaviour of trade union leaders, which will be discussed in chapters 5 and 6.

Trade union means organised labour, and organised labour means smooth running of the organisation. So employers should not be suspicious about the union activity as it was in the 17th and 18th centuries. On the other hand, trade union leaders should expect organised labour for better handling of the labourers and union activity as well. So both the parties should behave sensibly and cooperatively for the better performance of trade union movement.

One of the most important functions of trade union is collective bargaining. There are three tiers in negotiating the labour dispute between and among the concerned parties i.e., trade union, the management and Government. At plant level, there exists bipartite negotiation between trade union and the management. The collective bargaining agent (CBA) on behalf of the workers must have registration according to the ordinance (Sec. 7, IRO, 1969). If there are more than one registered trade unions in an establishment, then the CBA is determined through secret ballot organised by the Labour Department if any of the unions or the employers apply for it.

In the early part of 1984 thirteen national federations united to form a loose confederation called Sramik Karmachari Oikya Parishad (SKOP), an alliance of workers and employees. Subsequently, by early 1990s, the numbers of such federations were seventeen. Most of them belong to 23 registered national federations. (Hussain et al, 1998: 160). The leaders of the Parishad have placed a five point demand programme which includes free trade union activities, stopping further denationalisation of industries, dearness allowance of 30 per cent to all employees and workers irrespective of private or public sector, immediate publication of industrial workers wage commission report, setting mills and factories by reducing unproductive expenditure etc. The SKOP leaders took up the matter to the Government for realization of 5 points and arrived at an amicable settlement in July 1984. But it brought no result of the settlement (The Bangladesh Observer, January 7, 1985).

In many industries of Bangladesh two or more unions are in existence. There were only three unions in Bangladesh Railway in 1974. But now there are eight unions (Taher et al, 1998: 104). Multiplicity of trade unions may be well understood by the fact that at the national level there are as many as 26 national federations of trade unions, most of whom are attached to different political parties. More unions competing for membership within a single group is a regular feature in our manufacturing concerns. This leads to disunity, and disunity leads to fragmentation and consequent weakness of trade union movement. Multiplicity of trade unions creates rivalry among unions due to different political affiliation of union leaders. (Taher, et al, 1998: 103). These things happen in developed countries too. There were sharp ideological conflict between the communists and the non-communists in France. The trade unions, which command the widest support in France and Italy, are under communist leadership in sharp opposition to the prevailing political regimes. Non-communist faction was relatively small, but the Government had a favourable notion about them as they were not so much involved in politics as the British trade unions (Cole, 1962 : 32). In Holland, there were five rival trade union movements, working for the most part together on wage questions, but divided sharply by political and religious differences. In Belgium the main trade union body is integrally connected with the Belgium labour party, but there was keen rivalry with the Catholic trade unions, which were more loosely related to the Catholic party. In Australia there were sharp difference between communists and non-communists (Ibid, 3).

Sometimes regional feeling works which is detrimental to unity and integrity of trade union movement. Moreover labourers and even labour leaders lack in academic qualification and social competence as they belong to poor socio-economic status (Bhuyan, 1991: 156).

The common cause for clashes between unions at various plants in Bangladesh is the influence of outside leaders, but this practice has been started since the inception of the movement in this subcontinent. There were some genuine reasons behind the involvement of a group of leaders, commonly known as outsiders or professional trade unionists (outsiders means those leaders who are not employees). In the early days of factory system labourers were unorganised and helpless in front of capitalists and foreign Government, and they were the close allies. To keep contact with political parties and political personalities or participation of politicians in trade unions is not bad if it is not derogatory to the interest of the working class. But it became a fashion in India to condemn the participation of trade unions in politics and more emphatically the participation of politicians in trade union. In fact, trade unions in all countries are more or less political in the sense that they are affected by the decisions of the state and therefore try to affect those decisions (Crouch, 1979: 8-9).

Moreover they were mostly illiterate, low paid and not having any social status. In this situation prominent politicians came forward to help them to be organised, and encouraged them to form trade union. Workers will respond at the time of anti-British movement. The outside leaders came as the philanthropist and sympathisers, they did some potential but later on they became full-time organisers and political leaders (Rahman, 1969: 158).

During pre-liberation period outsiders who were not workers themselves ran majority of trade unions. These outsiders could be divided into three categories. In the first category there were persons who according to the strict legal sense, were not workers but were usually related to be industry in one way or the other. These individuals might be workers who were discharged for trade union activities or those who after becoming active trade unionists could not afford to work in any establishment. Generally, such people turned to professional trade union leaders. The second category included other professional leaders who did not belong to the working class but had adopted trade unionism as a career. In the third category, there were workers of communist and other political parties who joined the trade union movement with the main purpose of establishing the political base (Ahmad, 1978: 39).

But too much politicisation hampers the ultimate goal, i.e., interest of the workers on one hand and it helps the political leaders to exploit the union on the other. The trade unions of most of the developing countries are now involved in politics to such an extent that often seems to obscure the economic interest of its members (Millen, 1967: 1).

In underdeveloped countries especially in Asian countries leadership of the trade union organisation is a very mixed grill. The leaders tend to be an elite group within their sphere, and specially in Asia, to be outsider to the industrial framework. Leadership that developed within the working classes of course exception in some places of Asia. For example, sharp intelligence and high quality leadership made Mr. P.P. Narayanan, a Malayan plantation worker and Secretary General, Malayan Plantation Workers Union, an elite despite his limited formal education (Ibid, 27). Mr. H.R. Chowdhury, a plantation clerk was then financial secretary of the same union having enough academic background.

In most of the Asian countries professional people like doctors, teachers, civil servants, most frequently lawyers devote either part or full time to the task of trade union leadership. Especially in India and Ceylon such professional people, frequently of high caste have assumed direct leadership in trade union activities. For example, Mrs. Maniben Kera, President of the 40,000 members of Western Railway Employees' union in India, a Social Welfare oriented 'outside' leader had given long valuable service to the trade union movement. Mrs. Boze was professionally a physician in Calcutta port and a labour leader as well. She was a leader of Calcutta dockworkers union and in 1962, elected President of INTUC. Renowned labour leader and an author, S.A. Dange, Secretary General of the AITUC, came from a elite family from Bombay. Many other leaders came from such distinguished background (Ibid, 27). Not only the professional people, prominent politicians and eminent personalities joined the trade unions in Indian subcontinent. Personalities like Mahatma Gandhi, V.V. Giri, M.N. Roy, Jai Prakash Narayan, Gulzarilal Nanda, Hussain Shahid Suhrawardy, Abid Ali, Ashok Mehta, Deben Sen, Kumruddin Ahmad, Dr. M.A. Malek, Mrinal Kanit Bose etc. provided able leadership to the trade union movement.

In Singapore, professional people operated from behind the scene. They used to advise the unions, mostly they were political figures and they had the aim to influence and guide them politically. They might be outsiders but extremely dedicated and capable leaders (Ibid, 27). Mr. N.M. Perrera, a graduate of the London School of Economics who was the President of Ceylon Federation of Labour and one of the major port workers unions. Bala Thampose, a well-known lawyer, who also represented and influenced other unions, led the Independent Ceylon Mercantile Union. A former middle-level civil servant led the Government Clerical Service Union. The Ceylon Communist Party was headed by Peter Keuneman, an economist, who was also the leader of the local Communist Party (Ibid., 1967: 27-28).

In the case of Pakistan, Dr. A.M. Malik, a physician became a professional leader in labour front. Eastern Pakistan Trade Union Federation (EPTUF) was formed on the 28th of September 1947, the first and most representative organization of the workers in the then East Bengal. Dr. A.M. Malik decorated the post of President, Faiz Ahmad as Secretary, Nepal Nag, Mohammad Ismail and Mohan Jamadar as Vice-President and Anil Mukherjee and Gour Barman as Assistant Secretaries. Most of the top rankings were professional leaders. Jogendra Nath Mondal was the first labour minister of Pakistan and Dr. A.M. Malik was the labour minister in the East Bengal cabinet of Khawaja Nazimuddin. Later on when J.N. Mondal left Pakistan permanently Dr. Malik was sworn in as Central Labour Minister (Ahmad, 1978 : 32-34).

Another important professional leader was Kamruddin Ahmad. He was associated with labour movement since 1945. He was elected President of the Post and Telegraph Union (Dhaka Branch) and R.M.S. employees' union, Dhaka in 1945. In 1947, after independence he was elected President of the East Pakistan Railway Employees' League, Dhaka Branch in 1948 and elected Vice-President of the All Pakistan Trade Union Federation (APTUF) he was also elected President of the East Pakistan Federation of Labour in 1950 and continued as its President until 1955. He attended tripartite labour conference and I.L.O. meeting in Havana and Geneva. He was a politician and eminent author and joined the Diplomatic Service of Pakistan in 1956 too.

There were other professional leaders, e.g., Faiz Ahmad, S.M. Sulaiman, B.A. Bakhtiar, Chowdhury Rahamatullah, Abu Syed Anwar, M. Suleman (Chandpur), Aftab Ali, Khatib, Mirza Ibrahim, A.R. Sunnamat, Dinen Sen, Samar Ghosh, Gholam Murtaza, Mahbubul Haque, Umar-Din, Nurul Huda, Aswani Dev, Anil Basak, Fazal Elahi Qurban, Ramesh Chandra, Narayan Das Bacher, J. Bukhari, M.A. Khan, Prof. Sibdas Ganguly, Abdul Ghafur, C.P. Dave, Mubarak Sagir of both East and West Pakistan. These people contributed a lot to the development of trade union movement of the country. Some political leaders used to have close contact with the workers to mobilize support of the workers for their party. Abdur Samad of Awami League and Muhammad Toha, both members of East Pakistan Assembly were entrusted with the task of organising labour movement. Mirza Ibrahim was working in close collaboration with National Awami Party (NAP), good organiser in the labour field. In East Pakistan NAP had organised a labour front in the name of Pakistan Mazdoor Federation (Ahmad, 1978: 40).

The story of Bangladesh is also the same. Mrs. Shahjahan, Saifuddin Manik, Aftab Ali, Qazi Zafar Ahmad, Abdul Mannan, Oli Ahad, Ruhul Amin Bhuiyan, Haider Akbar Khan Rano, Sirajul Hussain Khan, Maulana Saidar Rahman, A.B.M. Mohiuddin, S.K. Pal, Dewan Sirajul Hussain, Ashraf Hussain are the leading union leaders of Bangladesh. They mostly do not belong to the working class. But they have adopted trade unionism as career which is not congenial for the growth of the movement. Still it is in practice in Bangladesh. As the very few of the workers have the training, education and information which would enable them to bargain effectively with employers of private sectors and bureaucrats in public sector corporation (Ismail *et al,* 145). However, in order to overcome the problem, the industrial policy of 1991 suggested that the office bearer of trade unions can in no way become persons other than factory workers and from political influence. So it clearly indicate to disassociate workers from politics (Mondal, 1992: 17). In India, legislation was reportedly pending in 1962 that would prohibit outside leadership among Government employee unions; in Pakistan, outside leadership has been limited to 25 percent of any given union (Millen, 1967: 28).

So the practice of becoming leaders from outside the industry was a common affair since the very beginning. Since the outsiders controlled the unions, the leaders did have little knowledge regarding the plights and the conditions of the workers and little they could do for the betterment of the workers. What actually they tried to do was that they only tried to fulfill their own interests while the interest of the working class were almost neglected.

Political involvement of unions is found not only in Bangladesh; but it is prevalent more or less in other countries of the world also. Though politics is the secondary activity, when they fail to protect the vocational interest of its members by negotiation and consultation, the unions enter into politics (Roberts, 1962: 67).

The following are the common causes of conflict between unions at plant level in most of the cases:

(a) Multiplicity of plant unions and rivalry amongst them appear to the result of management's orthodox notion of keeping its working class under control if necessary by favoured treatment to one group of workers and their leaders over others.

(b) Heterogeneous working class population.

(c) Influence of outside leaders.

(d) Arrogant attitude of trade union leaders and workers (Rahman, 1968: 91-92).

Though the IRO 1969 has restricted the outside leaders, to be the officials of basic trade unions especially the different political parties have a definite impact on multiplicity of trade unions, which leads towards rivalry of unions in plant level and in the country as well. The rank and file leaders of basic trade unions were under the direct guidance of different labour leaders of different political parties and professional trade unionists under the erstwhile Act.

The ruling party extends support to a particular working group to achieve their political ends, which makes the inside leaders of basic unions to be oriented to political party in power. This is the practice of almost all the parties in power especially in developing countries like Bangladesh. Thus the most leaders of basic unions have affiliation to different political parties. This unfair practice makes the union leaders corrupt, disobedient and unsullied as they have support of the party in power. Then they become more interested and active in extending their political base over large membership rather than mitigating problems of the general workers (Taher, *et al,* 98:99). So there is clear-cut division between the trade unions backed by the management and ruling party and trade union with opposite parties. This vitiates the atmosphere for the healthy growth and development of trade unions. This vicious circle of political unionism, multiplicity and rivalry among the trade unions, low membership, unsound finance, lack of welfare work, weak collective bargaining, preoccupation with litigation and strike, victimization, short falls of labour laws are the major bottlenecks of trade union activities in Bangladesh. Here the majority of the unions are very small sized and plant based. Not only in post-liberation period the design has also been in practice from Pakistan period (Rahman, 1969: 33). Rahman says that most of the unions in Pakistan have so far been organised on the basis of 'Plant' or 'establishment'. It may be termed as 'plant unions' or 'house unions' rather than industrial unions proper (Rahman, p. 33). The multiplicity of unions which are financially weak and managed very poorly, conflict of leadership in particular plant and national level also cause serious problem to develop a strong trade union movement in Bangladesh. The growth of the number of registered trade unions and their members in Bangladesh is shown in the following table.

From table 1.1 below, it is evident that there has been a rapid increase in the number of registered trade unions as well as in terms of number of members of these unions.

Table 1.1 : Growth of Registered Trade Unions and their Members

Year	*Total Number of registered trade unions*	*Total Number of members of registered trade unions*	*Increase in number of registered trade unions %*	*Increase of the number of members %*
1949	74	98543	100	100
1949-71	1160	450606	1567.57	457.27
1971-97	5319	1793074	458.33	397.23
1947-97	-	-	7187.84	1819.59

Source : Department of Labour, Government of Bangladesh, 1999.

Objectives of the Study

The broad purpose of the present study is to investigate the characteristics of the trade union activities in Bangladesh during 1947-2000. The specific objectives of the study are as follows:

(a) To identify and assess the impact of trade union activities on production and productivity and on management performances.

(b) To analyse the development and performance of trade union movements in Bangladesh during the period 1947-2000.

(c) To identify the socio-economic development of industrial workers and their participation in trade union activities.

(d) To analyse the growth and development of trade unions and their membership in jute mills of Bangladesh during the period.

(e) To assess the impact of trade union on production and productivity of jute mills in Bangladesh.

(f) To evaluate the socio-economic background of trade union members vis-à-vis their participation in trade union activities as those affect employees' morale, attitude, motivation and job satisfaction in jute mills.

(g) To assess the management performances of the Jute mills and locate the impact of trade union activities on it during 1947-2000.

(h) To pinpoint basic managerial efficiency of the Jute mills in Bangladesh and its relevance to trade union activities.

Thus the purpose of our investigation is to evaluate the attitude of workers towards trade union workers and trade union leaders. Trade union leaders' experience with the management and the toiling masses has come out through the exclusive interview with them. Finally, on the basis of data collected, we have to assess the performance of trade union activity on the jute mills of Bangladesh.

Research Hypotheses

The following hypotheses have been formulated out of the extensive review of relevant research studies done earlier in the context of the trade union activities in Bangladesh.

1. The first hypothesis is that we would like to test is that there is a positive correlation between trade union activities and productivity of the jute mills. This suggests that trade unions can better help the management if the management is labour friendly. This enhances the level of production of the respective jute mills. We would like to test this hypothesis taking data from the two private sector jute mills. The testing of this hypothesis is purely data based and does not depend on the wishes of the researcher.
2. The second hypothesis is that there is significant variation in socio-economic development of workers of the jute mills due to variation in trade union activities among the jute mill's in Bangladesh. This hypothesis is again based on the data collected for our purpose.
3. The third hypothesis is that employees' morale, motivation and job satisfaction are closely related to trade union activities. The testing of this hypothesis again is based on the data collected from field survey.

Methodology of the Study

As is evident from above, the present study is, in part, based on the existing literature and secondary data, and in part, on field survey. To find out the trends in the performance of trade union activities and its impact on industrial production, productivity, managerial efficiency and socio-economic development of industrial workers with special focus on jute industry in Bangladesh, we shall draw on the existing literature and secondary data. For secondary data we have heavily relied on the available of books, periodicals, govt. reports, Bangladesh Jute Marketing Corporations, the Bureau of Business Research, Bangladesh etc. For the literature and secondary data we have used the resources of the library of the University of North Bengal, the National Library, Calcutta, Library of the V.V. Giri National Labour, Institute, Noida, U.P.; Bangladesh National Library, Dhaka; Divisional Library, Rajshahi; Rajshahi College Library; Library of the Rajshahi University and Varendra

Research Museum Library, Rajshahi; Bangladesh Development Studies Library, Dakha; ILO Library, Dhaka and Bangladesh Institute of Labour Studies (BILS).

As regards the methodology of the survey of trade union members and officials, the detailed method is out lined below:

(i) ***Universe of Study*** **:** The universe of study is the whole of Bangladesh.

(ii) ***Sampling Frame*** **:** Three types of sampling have been done, viz., the stratified, the purposive and the random sampling.

(iii) ***Sampling Procedure*** **:** Field survey has been conducted through three sets of questionnaires. The first set has been used for trade union members. Data have been collected from the plant level managerial personnel through the second set of questionnaire. Third set of questionnaire has been utilized to gather information from the trade union leaders of the respective jute industries.

For different types of information we have collected materials from four jute mills. Of these two jute mills are privately owned and managed and the rest two are nationalized. The private sector jute mills that we have selected are AJAX Jute Mills Ltd. established in the year of 1965 and Mohsin Jute Mills Ltd. established in 1969. Comparatively AJAX Jute industry is bigger than the Mohsin jute industry in terms of production units and the employment of labourers. The AJAX jute industry is located in Mirer Danga, Daulatpur, Khulna of Bangladesh and also the Mohsin jute industry is in Siromani, Khulna. The two nationalized jute industries are Jessore Jute Industries Ltd. established in 1970 and Rajshahi Jute Mills Ltd. (1968). Of these two jute mills the Jessore Jute industry is bigger than that of Rajshahi Jute industry in terms of capacity of production and the volume of workers engaged in the production. To select two jute mills from each sector (private as well as nationalized) we have first stratified the jut mills into two groups big and small. Obviously, we have purposively selected one bigger and one relatively smaller jute mill from the strata of privately managed group of mills. Similarly, one relatively bigger and one smaller jute mills have been chosen from the nationalized jute mills following the same procedure.

(iv) ***Tools and Techniques of Data Collection :*** Data have been collected by the researchers through questionnaire by personal interview with the relevant respondents.

(v) ***Data Processing :*** After the collection of data, these were processed through relevant statistical calculation manually by using calculators and also through computer. The tools used for analysing data are tabulation, simple frequency counts, simple averages, percentages and some statistical tests.

Review of Literature

Rahman (1967) in his unique study of late sixties on Organised labour and Politics in Pakistan contends that trade union movement is a much more accepted phenomenon today than it was in the 18th or early 19th century. We find friendly societies or associations catering to the economic and non-economic needs of a group of people having similar needs and problems, much before the appearance of the trade union in its modern meaning and form. Trade unions, as we understand them today, are however, a creation of the industrial revolution, which made direct contact between employee and the employer almost impossible. The author explained very critically the link between trade union and politics. Despite the link main objective of the paper is to present a picture of the political factors, in general, and the political happenings in particular, through which the movement had, and still is to, pass.

Rahman (1969) in his study on trade union Management and Leadership in Pakistan has stated that with the development of industry and commerce, the country saw more and more workers under the fold of trade unions. In any major industrial city of Pakistan people will come across trade unions that have been formed to look after the interest of the workers. In this article an attempt has been made to analyse how these unions are managed and what type of leadership is available for the management of these unions. According to Rahman efficient trade union management needs dedicated leaders, and these leaders should preferably come from the rank and file who are expected to be in better position to understand the problems of their fellow workers. The attitude of the employers should also be changed regarding the existence of the trade unions. Trade unions should be accepted as a reality and considered as a cooperating agent in the production process.

Rahman's (1969) paper throws some light on the structures of trade union organization in Pakistan. To make the discussion vivid and comprehensive he has touched upon the stature of unions prevailing in some developed countries of the West, especially in the United Kingdom and the United States of America. First he has discussed the types of trade unions. Then he has shown very nicely why craft union predominated for a long time in the West, specially in the United Kingdom, and in Pakistan the trend was towards industrial unionism. Most of the unions in Pakistan have so far been organized on the basis of plant or establishment. So these are plant unions, or house unions, rather than industrial unions in proper. Therefore most of the unions in Pakistan were very small, so their success or failure depended upon their own individual efforts. In many a firm there could be found more than one union, resulting in conflict and bitterness among the rank and file. The multiplicity of unions is accentuated by the existence of multiple industrial and national federations, which are guided, directly or indirectly, by influential political leaders.

Chowdhury (1983) in his study says that disputes are part and parcel of industrial life, which may arise out of differences among the actors in several matters prominent of which are substantive matters. There are also ways of handling disputes, which at time may be at variance with the state. These and other related issues have been discussed in this paper.

Alam (1983) in his study has recognised the hard fact that disputes can never be avoided, they can only be resolved as and when they occur. Different methods have been evolved over the course of time for settling industrial disputes and different countries have applied them in varying degrees. The methods used for the settlement of industrial disputes in Bangladesh are collective bargaining, conciliation, arbitration and adjudication. The present form of many of the legal provisions concerning the aforesaid methods is practically the sequel of many ups and downs that occurred in the past in the Government labour policies and labour laws. The study, therefore, mainly attempts at highlighting and depicting the various legal provisions on settlement of industrial disputes as per their development perspectives.

Bhattacharjee (1985) has made an attempt to examine some theories of industrial unrest, focus on the dynamics of unrest in selected South Asian countries and identify some reasons pertaining to unrest. Available data show that the phenomenon of unrest has been common to India, Pakistan, Bangladesh and Sri Lank under review. There were variations in the intensity and magnitude of industrial unrest from year to year in these countries. The causewise analysis of unrest in the country indicates that wages was one of the factors influencing unrest, but this was not the only factor responsible for the same. The study shows that the issue of wages is one of several factors contributing to industrial unrest, but neither the consumer price indices nor the change in the real wages substantially influences the intensity of labour unrest in the given labour surplus economies. The variation in the intensity of unrest depends on the complex set of socio-economic and political factors.

Saha (1986) in his study has tried to present two things very nicely. First, research has been made so far in the field of raw jute and jute manufacturing industry of Bangladesh and the second, the development of jute industry with its different problems and prospects. The Research on jute was started in the year 1900 at Dhaka with the appointment of the fiber expert committee on the recommendation of the Royal commission on agriculture. The jute research institute was established at Tejgaon, Dhaka in 1957. Since then good number of research has been made on raw jute and matters relating to jute industry of Bangladesh. The object of the present study is to evaluate the most noteworthy studies to arrive at a definite conclusion about the major problems on the survival of the jute industry in keen competition with the synthetics substitutes and other jute goods exporting countries.

Khan (1986) in his paper examines the attitude of the Government towards industrial workers through a historical analysis of all the labour policies declared so far basing on the political and social changes that occurred in the Bangladesh territory. This analysis has been made over three distinct time spans, e.g. the British, the Pakistan and the Bangladesh periods. Having summarised the declared labour policies of the Government in each time phase the paper has attempted to identify the extent to which those policies have been related with the labour laws for implementation. It has been pointed out in this paper that while there was no formal policy regarding labour in the British period, the whole of Pakistan and Bangladesh periods labour policies have been declared in different times. But very few of those were in fact implemented in truest form.

Siddique (1989) tried to identify the main factors that had contributed to the development of the industrial relations system in colonial India. Special attention has been given to the role of the colonial Government in the development of the industrial relations system in India. The author has focused more in this paper on the issue that the colonial Government had followed a policy which would have best served the interest of the 'home' country, i.e., the imperial power in setting the industrial relation system in colonial India. Moreover, he has mentioned that most of the early capitalists in colonial period were 'foreigners', who belonged to the ruling class. As a result, the relationship between the workers and the owners of the industries was one of 'servant-master' rather than employer-employee.

Khan (1990) believes that under the system of modern management the relationship between trade union leaders and management is important. Such relationship stands as a pre-condition to the advance of harmonious industrial relations in general and trade unionism in particular. In a backward and poor country like Bangladesh, ruled by undemocratic regimes for a long time, such precondition is absent. Absence of democratic practice, both at the enterprise and national levels, leaves little scope for trade union leaders to maintain close or links either with the rank and file the management. The often-used practices of victimizing and/or bribing popular leaders of trade unions have rendered most of them quite ineffective to their ultimate cause. So a true democratic environment is required for the emergence of right type of trade union leadership.

Bhuyan (1991) holds the view that the healthy growth of trade union movement in Bangladesh is faced with numerous problems created by the successive regimes not only through promulgation of various ordinances and undemocratic labour laws but also by forcible occupation of the existing trade unions in Mills and factories. According to Bhuyan, the healthy growth of a democratic movement depends mainly on two factors: (a) internal strength of the trade union organisation and (b) external environment favorable to trade unionism. Internally, trade union organisation and movement in Bangladesh suffers from many inherent weaknesses or limitations. Most of the workers are illiterate, unorganized and of poor background. Approximately 8% of our workforce is employed in the organised sector of our economy. The majority of the workers are engaged in agricultural sector. They are mostly unorganised or organisationally weak and so poor that they cannot even protest against their employer and the exploitive socio-economic structure of the society as a whole. Disunity and disintegration is the main problem of our trade union movement. Disunity cannot be wiped out so long as inequality of income and wages exist between different categories of workers working in organised and unorganised sectors of the country.

Chakma (1991) in his study has tried to examine the validity, in the context of Bangladesh, of the popularly held belief that workers in the developing countries lack commitment of industry. His study finds that workers are fairly committed to industry. If we take the respondents of this study as representative of the total industrial labour force of Bangladesh, we may say that most of the workers in Bangladesh are either highly or moderately committed to industry. Workers' socio-personal factors like age, education, rural urban background, nature of employment, occupational background, length of industrial work-experience, and family attachment, have significant relationship with commitment. Factors like marital status and dependency load, however, have nothing to do with workers commitment. The author has pointed out a very interesting observation that there has been a negative relationship between education (level) and commitment. The ideally committed industrial worker in Bangladesh a bit older, has had little or no education, was born and brought up in an urban environment, has a background of urban or industrial occupation, has a fairly long industrial work experience and lives along with his family members in and around the place of work.

Hussain and Haque (1991) investigated the socio-economic background and characteristics of trade union leaders, particularly in the cotton textile industries in Bangladesh. The author has made an attempt also to see their political liabilities as well as their honesty and sincerity to union activities. The study reveals that there is significant influence of age, experience, skill, education, marital status, family size, locality and political abilities on union leadership at the basic union level. However, it shows no significant impact of training on union leadership. Most of the union leaders at the enterprise level assume leadership without much educational preparation in Bangladesh. It is found that most of the union leaders have direct involvement with political parties and they lack sincerity and honesty to the interest of the workers and enterprise as well. The above picture of the union leadership, though cannot be treated as a complete one of the country, appears to be frustrative no doubt and need taking early steps to check the further deterioration, so that unfailing remedial measures may be taken to improve the qualities of union leadership at the enterprise level and thereby ensuring an effective growth of strong and viable trade union leadership in the country.

Khuda and Barkat (1992) have shown in their paper that the labour force in Bangladesh has been rising from 9.6 million in 1901 to 50.7 million in 1987. This trend of accession of the labour force varies in case of male and female in the said period. The male labour force increased by over three times during the 1901-89 period, whereas the female labour force registered an unbelievably high growth rate in 1989. The very unprecedented increase in female labour force in 1989 is primarily due to a change of definition of what constitute an "economic activity". Furthermore, the problem of over reporting cannot be ruled out. In comparison to other countries, the female labour force participation rates at the younger age are quite high in Bangladesh. This signifies a sheer wastage of potential quality of manpower, necessary for economic development of the country. The increased rate of females has not been pronounced in any industrial group, except manufacturing. Much of the increase in manufacturing has resulted from increased female employment in garment industries. A handsome number of the non-agricultural labour engaged in Bangladesh is still traditional. This is especially true in case of females, large portion of whom are engaged in domestic services.

Karmaker (1992) in his paper has explained and identified possible measures to improve the system of the quality of working life in the industrial sector of Bangladesh. The Quality of Working Life in an industrial sector has been assumed to be the result of a complex set of three components viz., industrial welfare measures, industrial democracy and wages related measures. The QWL has received enthusiastic support from managers, workers and government agencies in both developed and developing countries. The issue of improving QWL is yet to draw the due attention of the concerned agencies in Bangladesh. This paper also aims at reviewing the indicators for measuring QWL, evaluating the same in the industrial sector, and identifying possible measures to improve the system of the QWL in the said sector of Bangladesh. According to the author, to improve the issue of QWL, interactions amongst the parties are very important. Concerned parties should sit together to ensure that the decisions of the sitting or meeting are given effect to in the spirit of tripartism. Tripartite actions are very much essential to make the programme a success. It is clear from the study that the QWL, among other factors, is directly associated with minimum wages guaranteed to all working people irrespective of occupations taking cost of living into consideration.

Khan (1992) opines that the term 'conflict' is very common in industrial arena. It is a necessary evil of an industrial society; rather we can say industrial conflict is a two-sided coin. Modern researchers believe that just as the level of conflict may be too high and require a reduction, it may also often be too low and in need of increased intensity. A moderate amount of disputes, if managed properly, may have a positive contribution to productivity. The author says that too much industrial dispute is detrimental to social and economic progress, but too little of it may lead to stagnation.

From this study it is revealed that post-liberation period in Bangladesh was found to be more conflictual than that of the pre-liberation period. Turning exclusively to the post-liberation era, a comparative analysis was made to see the trends of industrial disputes over different ruling regimes. Each consecutive regime of the post-independence Bangladesh was found to be progressively conflictual over the previous one. That meant that Bangladesh Nationalist Party (BNP) regime (1976-81) was more intense in disputes than the Awami League (AL) regime (1972-75) and the Jatiyata Party (JP) regime (1982-89) was more intense than the BNP regime. Dispute patterns and intensities during the present democratic regime could not be examined due to its shorter duration on the one hand and non-availability of data on the other.

The author explained the differences of intensity of conflict in different regimes as follows:

The AL Regime was the first regime immediately after liberation. Sheikh Mujib, the then head of the regime, declared that it would not be possible for the government to give anything for the people for the next three years, and in line with the declaration, workers were also mentally prepared not to manifest some of their demands as a price for independence. National Emergency was also in force in 1974. All these factors might have accounted for the relatively lower intensity of disputes in Bangladesh during the A.L. regime. Worsened economic condition in the past coupled with political liberation might have accentuated the dispute intensities in the BNP regime. The highest intensities of disputes in the JP regime might be explained in terms of workers protests against martial law and its restrictive measures in the initial years, country wide movement for workers' interest led by SKOP (Sramik Karmacheri Olikay Parisad) and frequent political movements in consonance with strikes and cartels in mills and factories in demand for the removal of autocracy.

Karmaker (1992) has done a study on factors influencing Industrial Relations System (IRS). The author first defines the industrial relations system concept. Next he opines that most of the studies have been done in the context of highly industrialised nations. The review of these studies indicates that the problems were approached mostly ignoring necessary interconnected components of the system of industrial relations operating in developing nations. Industrial relations include not only labour management relations within the plant, but also those relationship in community and society, which grow out of the labour management relationship. In fact, the climate for optimum industrial relations depend on interaction of employers, employers and state whose behavioral patterns are governed by a number of inter connected and interdependent variables. Thus, a broader framework of industrial relations is the complex of influences, which play their role in setting the pattern of industrial relations system.

Saha (1993) in his paper presents a comparative analysis of human resource management (HRM) practices between China and the West (i.e., primarily Canada and the United States). It contends that research into international HRM practices should incorporate the combination of work environment and organisational characteristics of target countries in order to ensure a more complete understanding of the similarities and differences that may exist between countries. The author feels that though human resource management is a well-organised field of study, it is not the case in China. Very little is known about Chinese management of human resources. The analysis of Chinese HRM practices that has been presented in this paper required more than two years of information collection from diverse sources. China is a vast country and its linguistic and cultural diversity makes it a difficult subject for investigation. Despite so many limitations the author presented a scholarly article of worth reading for the fellow scholars.

Khan (1993) in his theoretical analysis has shown that every industrial conflict may have unique dimensions of its own, and as such its resolution may be subject to some contingent conditions which may very in number as well as in magnitudes from one case to another. Basing on the available literature on the subject, this paper attempts to make a theoretical analysis of these contingent variables. It has argued that once one is well aware about these conditions, one should be able to properly assess the conflict situations and then use appropriate institutional methods aimed at their constructive resolution.

Taher's (1997) study attempts to examine the conciliation machinery (CM) in Bangladesh. The findings of the study, which are based, both on primary and secondary data, reveal that although the performance of CM is disappointing, it played an effective role in the dispute settlement. Though the CM plays a very effective role in the settlement of industrial disputes, the failure ratio (i.e. type of settlement made in conciliation) of the conciliation sessions has increased enormously that leads to bitterness among the concerned parties this is damaging industrial peace, progress and productivity. Finally the paper has given some suggestions for improving the effectiveness of CM in Bangladesh.

Moniruzzaman and Parveen (1997) in their study on the functioning and effectiveness of collective bargaining in Bangladesh have recognited collective bargaining as an essential component of its industrial relation machinery from the very beginning of industrialisation. Though the contribution of collective bargaining process has been significant in the country, it's functioning and effectiveness is still far from exception. The following factors are responsible for poor functioning of collective bargaining: (a) Lack of proper education and training of bargaining agents, (b) multiple trade unionism, (c) inter union rivalries and (d) politicalisation of trade unionism. The study suggests some steps, which may improve the present state of collective bargaining, thus paving an avenue for industrial peace.

Nurullah (1985) has done one study on the "Pattern of Trade Unions in Bangladesh". The study analyses the organizational pattern of trade unions in Bangladesh. The study specially focuses on the size, concentration, type, management leadership, negotiation, relations to other unions and finance at various levels. He has also made an attempt to discuss the legal aspect of a trade union organisation. The author has found that the majority of the unions in Bangladesh are very small sized and plant based. This pattern of union organization is, to some extent, responsible for the multiplicity of unions which are financially weak, poorly managed and above all, inclined to be biased and influenced by outsiders, those who have no relation with the working people.

Nurullah (1995) in his other study looks into the industrial relations system in the public sector textile enterprises in Bangladesh focusing on some important formal issues: collective bargaining, dispute settlement and grievance handling, labour management participation and role of labour welfare/employee relations service. The author also attempts to study legal aspect of the issues in shaping the industrial relation system. It is revealed from the study that the working condition at the enterprises under the Bangladesh Textile Mills Corporation (BTMC) and the legal provisions of labour welfare were not being propertly maintained. The study of industrial relations system in the public sector textile enterprises further reveals a conflicting relations environment. Though there was legally limited scope for collective bargaining, it happened that BTMC management negotiated bargainable issues and entered into agreement with trade union federations. The Government mediation was frequently used in setting disputes between the corporation and trade unions where industrial action was prohibited.

Sengupta and Moniruzzaman (1997) in their study have made an attempt to review the trends, causes and impact of industrial disputes in Bangladesh. The important feature of the industrial relations in this country revealed by data is that no correlations exist between the figures relating to the number of disputes, workers involved and man-days lost due to these disputes. The main causes of disputes are wages, bonus, poor working condition, suspension and dismissals, even retrenchment. A production and wage loss due to dispute is not ignorable as Bangladesh is a developing country with a low G.D.P growth rate.

Taher *et al.* (1998) in their study have found that multiplicity of trade unions (MTU) is a normal phenomenon of industrial society of Bangladesh. It exercises significant influence on industrial relation. This study indicates that MTU is the outcome of excessive interference by political parties who are interested to increase their influence in the area of trade union politics. It is also evident that most of the trade union leaders try to perpetuate leadership by virtue of their strong political affiliation particularly ruling party without any approval or consent of workers. Contemporary situations prevailing in the field of trade unions in Bangladesh, reveal the fact that MTU is detrimental to the interest of the workers. Finally, the study suggests some measures for nourishment and flourishment of trade unions in the truest sense of the term.

Fryxell and Gordon (1998) have examined the extent to which work place justice and job satisfaction predicted (happened to) with union management. They defined work place justice in terms of the procedural and distributive justice afforded by an organisation service system and, more generally, in terms of beliefs about a moral order in the work place. Multivariate analyses were used to test a number of hypotheses about the relationship among measures of institutional satisfaction, job satisfaction, and work place justice. The amounts of procedural and distributive justice afforded by a grievance system were the strongest predicators of satisfaction with a union, whereas belief in a moral order at a work place was the strongest predicator or satisfaction with management.

Jose (2000) in his unique study has made some important observations regarding the future of labour movement in developing countries. The changing socio-economic and political environment of developing countries necessitated the adoption of new approaches and strategies on the part of the trade unions to make them able to contribute substantially towards dynamic and equitable growth. He has presented the on going trend towards liberalization and globalisation which has thrown up new challenges to unions especially in developing countries for playing a meaningful role in society as well as opportunities for the same. In this situation the author has prescribed a revision of the traditional role of trade unions, the impact of changing work environment on unions and their responses to the same, with special emphasis on the different socio-economic, political and geographic settings of unions all over the world. The author further highlighted some key issues relevant to the future labour movements with particular reference to developing countries world over.

The author has made an attempt in this paper concerning the priorities of the labour movement in developing societies, which may be summed up as follows. A politically important option in the years to come will be to build on its established role in safeguarding social cohesion. This implies a strategic orientation to the long term goals of security, equity and justice for all in the labour world. The above goals can be achieved through redistributive transfer, specially aimed at correcting income inequalities and raising the level of social consumption and standard of living. All the strategies for achieving the goals need to be anchored in the mobilisation of diverse nature of interest groups in society on a political tent. An enduring position for the labour movement in developing societies means a relentless pursuit of the redistributive agenda.

Ali and Shams (2000) have done a study on the impact of trade unionism on wage, employment and work environment in the manufacturing sector in Bangladesh. Existence and nature of association between wage-labour productivity and employment labour productivity relation in an industry, on one hand, and nature of firms (unionised or not, state owned or private) on the other have been examined. The role of Government and political parties has also been considered. This paper suggests on the basis of the findings that trade unionism, in general has no positive impact on wage and employment in the manufacturing sector. Some apparently positive effect of trade unionism that was discerned in state owned enterprises appears to have waned in the recent past. Role of trade unions in creating anti-production environment seems to have been exaggerated. This paper also suggests that political parties in power and parties in opposition have to take their share of balance for the industrial unrest in Bangladesh.

Mishra (2001) opines that globalization is coming up as a concept for opening up of the domestic economy and intergrating it with the global economy. While the ongoing process of globalisation has created enormous opportunities in terms of accelerating the flow of international trade and investment as well as a rapid increase in the pace and scale of information exchange resulting in the emergence of a global market for investment finance, it has thrown up daunting challenges essentially arising from the denial of benefits of globalization to large sections of people and regions leading to gross inequities in the access to the emerging opportunities. This study discusses the social and economic implications of globalization and maintains that this is coexistent with the preservation of the rights of labour since a healthy and productive work force is a prime requisite for ensuring a strong and competitive economy.

Some important books reviewed

S.G. Panadikar's (1933) book *Industrial Labour in India* is a valuable and authentic book on labour in India. He has discussed the progress of industries in India in brief. Demand and supply of industrial labour in India have been discussed with due attention with the progress of industries. He have given special attention on labour in textile factories, in transport sector, mines and plantation. With conditions of employment. The author has given a special focus on the migration of labour from one state to another with its economic and social implications. There is a brief history of protective labour legislation and the programme of further protective labour legislation. He has also discussed the trade union movement, wages, health-hygiene, welfare and the industrial relations prevailing in India at that time.

In *Trade Unionism and Labour Disputes in India* (1978) Ahmad has presented a brief history of trade union movement in India from the very beginning with a background of agricultural labour. He has depicted the history of conflict between capital and labour in detail. Laws relating to trade unions and their implications and the right to combine have been discussed properly. The process of investigation and settlement of disputes have taken into discussion with due attention. Special attention has been given on the Ahmedabad Textile Labour Association. Moreover there is a special fours on different Acts related to labour and their interests.

V.V. Giri in his book, *Labour Problems in Indian Industry* (1962) has depicted the historical background of trade union movement. Diferent stages of development and salient features of Trade Unionism in India, industrial relations, system and reconciliation machinery for the settlement of industrial disputes and the labour legislations passed in different times and their implementation in industrial sector have been discussed in this book. Further he has given a stress on rationalisation and need for rationalisation in India, national wage policy, minimum wage living wage and fair wage, role of ILO regarding wage and social security package, industrial safety, health, hygiene and welfare in India.

In *An Introduction to Trade Unionism,* Cole (1962) has presented conceptual framework of Trade Union and their organisation specially focusing the British context with modern world as well. He has discussed carefully the type of trade unions, effectiveness of collective bargaining, legal aspect of trade union and the trade union involvement in politics.

B.C. Robert's book *Trade Unions In a Free Society* (1962) studies the organisation of labour in Britain and the U.S.A. in two parts, first one is union in Britain and second one is union in the U.S.A. The book is well written on British and American trade union scenario. In the first part, he has discussed trade union and wage problems, trade union and politics, trade union structure and organisation in British context. In second part, he has analysed the structure of union organisation, union democracy, wage bargaining and the control of inflation, industrial relations and unions and politics in the U.S.A. We can have a clear picture of trade union activities of the U.K. and the U.S.A. from this book.

A.S. Mathur and J.S. Mathur in their book *Trade Union Movement in India* (1962) depict nicely the evolution of trade union movement in India. They have discussed very systematically the origin and development of trade union movement in India and the obstacles to the growth of trade unionism with the extent of unionism. Moreover in a short span the authors have covered a vast country like India. Indian trade union structure, national dedications have also been discussed carefully. Laws relating to trade unions, source of trade union finance, collective bargaining method, legal enactments, trade union leadership, trade union and politics and some valuable observations have been portrayed vividly with due attention.

In the book *The Role of the Trade Union In Developing countries: A Study on India, Pakistan and Ceylon* (1963), Ronald D'Costa has presented a comparative discussion of trade unionism of East and West. He has especially tried to focus on the salient features of British and American trade unionism in brief. Not only he has pointed out the differences and similarities of Asian and Western unionism but also has discussed the trade unionism in the developing countries, specially India, Pakistan and Sri Lanka and some selected Asian Countries. Special attention has also been given on labour legislations, attitude of Government, Emplyers's attitude, collective bargaining and labour welfare activities.

V.B. Karnik's book *Indian Trade Unions : A Survey* (1966), is a unique work concentrating on India's trade union movement from its inception to mid-sixties of the last century, with problems and prospects. The author has specially highlighted the emergence of communists and the rift between communists and non-communists (nationalists) in the labour politics and national politics as well. He has also discussed the disunity and disintegrity within Indian trade union movements and later on the process of unity too. Second World War and its impact, post-war developments of labour movement had taken up very carefully.

Shiva Chandra Jha's (1970) book *The Indian Trade Union Movement* has dealt with very carefully the theories of Trade Union Movement, *e.g.,* Marxist theory and other theories. The other theories have been categorized as (a) British School—Sidney and Beatrice Webb, G.D.H. Cole, Harold J. Laski, (b) American School—John R. Comons, Selig Perlman, Robert F. Moxie, Frank Tarnenbaum, (c) Indian School—Mahatma Gandhi, Ashok Mehta. The author has also discussed the origin of capitalism and discussed trade unionism with different periods phase-wise with critical analysis.

G.K. Sharma's book *Labour Movement in India : Its Past and Present* (1971), has undertaken a tough task of explaining the origin, growth and transformation of the Indian working class. He has tried to focus the salient characteristics and peculiarities of labour movement in Indian. He has also given a detailed phase-wise picture of labour movement since 1875. He has also discussed trade unions in India since its inception to sixties of the last century with types of Indian trade unions, federation of trade unions and pattern of organization and structure of trade unions. Moreover the author has tried to analyse the relation between state and trade unions in India and in other countries.

Panchanan Saha has written a unique book title *History of Working Class Movement in Bengal* (1978). It is a beautiful work on the emergence of working class movement since the inception of organized sector of Bengal. He has also discussed the growth of organized trade union movement in India. Discussion on railway workers' struggle especially at Khagarpur and the general strike of jute workers have been focused in this book. Strike movements during 1936-39 and different aspects of working class movement during and after Second World War have been discussed in detail.

Kamruddin Ahmed's book *Labour Movement in Bangladesh* (1978) discusses how labour class has emerged with the development of plantation industry and other factory industry in Bangladesh. This is the only authentic book on labour movement in Bangladesh. He has discussed trade union movement of undivided India, Pakistan and Bangladesh chronologically. There are also analytical discussions on labour policy and labour legislations. He has discussed in detail the labour movement of Bangladesh with critical comments.

Harold Crouch (1979) in his book *Indian Working Class* (1979), has shown the relation between trade union and politics specially in the Indian context. He has discussed the relation between trade unions and the working class, trade unions and the state. There is also a valuable discussion on trade union movement of India with the inter-union rivalry, disunity in one hand and attempts on unity and co-operation on the other. He has also discussed the various aspects of trade unions and politics in India with special emphasis. In appendix II the author has given a brief life sketch of prominent trade unionists of India. It is really worth mentioning.

Ramanujam G. in his book *Indian Labour Movement* (1986) has narrated the emergence of factory system in India in a lucid language. The author, while tracing the growth of industry and industrial relations, as well as developments in political, economic and social life of India, has relatively given particular emphasis on the trade union movement of the country. He has analysed the growth of trade union movement of India with chronological documentation. He has also highlighted the role of Communist Party of Great Britain, the communist international and the Communists of the country in the growth and development of trade union movement. Special attention has also been given to security of employment, productivity, participative management, productivity linked wages labour legislations, social security etc.

Basudeb Sahoo's work *Labour Movement in India* (1999) is a unique work on problems and prospects of labour movement in India with special emphasis on Orissa. He has discussed carefully the concept of working class and changing profile of the workers of India as well. Thee is a beautiful discussion on Trade Union Movement of India and Orissa. There is also a discussion on changing pattern of women employment, problems of child labour and white-collar workers and their organisation.

Short Overview of Chapters

The overview of the chapters are as follows:

Chapter I : The Problem

Factory system emerged very late in India and Bengal as well in comparison to western world. So trade union activities started very late with obstinate resistance from both employer and Government in India and other Asian countries. These things have been discussed in this chapter in brief with the definition of trade union. We have also tried to pin point the problems of the trade union movement of Bangladesh. We have mentioned the broad purpose of the study and specific objectives of the study during 1947-2000. Some hypotheses have been drawn out of the extensive review of relevant research studies. We have tried to highlight the performance of trade union activities, its impact on industrial production, productivity, managerial efficiency and socio-economic development of industrial workers with special focus on jute industry of Bangladesh with the help of existing literature and secondary data collected from jute mills on the basis of open-ended questionnaire. Scheme of the work has been mentioned in brief.

Chapter II : Evolution of Trade Union Movement in the Indian Subcontinent During the British Rule

How labours as a class emerged in India and Bengal have been discussed with the gradual development of factory industry with a discussion of how trade union movement came up with lot of pressure from both employer and the Government. Trade union movement has been discussed phase-wise, *e.g.*, trade union movement since inception to 1920 to 1930, and from 1930 to 1947 respectively.

Chapter III : Growth of the Movement During 1947-1971

Trade union movement in the areas forming Pakistan in August 14, 1947 started under separate environment and Government. Most of the time the country was under military rule, where democratic environment had been threatened time and again. Under the circumstances trade union movement in Pakistan could not flourish in full swing up to 1971.

Chapter IV : Growth of Trade Union Movement in Independent Bangladesh (1972-2000)

This chapter deals with trade union movement of the sovereign country Bangladesh during 1972-2000. The chapter has been segregated : the movement from 1972-1981, the movement from 1982-1991 and the movement during 1992-2000.

Chapter V : Performance of Trade Union in Public Sector Jute Mills

In this chapter performance of trade union in public sector jute mills has been discussed carefully mainly on the basis of interview schedule for trade union members (workers), labour leaders and mill executives. On the basis of field data we have tried to evaluate the relation between trade union activities and industrial relation, trade union activities and productivity of jute mills and the relation between employees' moral, motivation and job satisfaction to trade union activities.

Chapter VI : Performance of Trade Union in Private Sector Jute Mills

In this chapter the nature of the study will be the same as in the fifth chapter. The only difference is that the study is concerned with two private sector mills, namely, the Ajax Jute Mills and the Mohsen Jute Mills.

Chapter VII : Summary and Concluding Remarks

This chapter summarises all the discussions in earlier chapters. Conclusions and policy prescriptions have been made finally for the growth of a healthy trade union movement in Bangladesh.

REFERENCES

Abdullah, Faruk, (1983), *Bangladesher Arthonaitik Itihash,* Dhaka University Grantha Sanstha.

Ahmad, Kamruddin, (1978), *Labour Movement in Bangladesh,* Inside Library, Dhaka.

Ahmad, Mukhtar, (1978), *Trade Unionism and Labour Dispute in India,* Longmans & Green Co. Ltd., Bombay.

Alam, Md. Fashiul, (1983), Industrial Disputes Laws in Bangladesh—A Study in their Development Perspectives, *The Cost and Management,* September-October.

Ali, Amin Muhammad and Shams, M. Nisar Ahmad, (2000), Trade Unionism in Bangladesh : A Study of the Impact of the Trade Unionism on Wage, Employment and Work Environment in Manufacturing Sector, *The Jahangirgar Economic Review,* Vol. 11, No. 1, June.

Bangladesh Observer, January 7, 1985.

Bhattacharjee, Durgadas, (1985), *Industrial Unrest in South Asian Region: A Focus on Theories and Trends.*

Bhuyan, M. Syefullah, (1991), The Trade Union Politics in Bangladesh: A Study of Textile Mills Workers' Union, *The Dhaka University Studies,* Part-C, Vol. 12 (1).

Bhuyan, Syefullah, (1991), A Study of Textile Workers' Union, *The Dhaka University Studies,* Part C, Vol. 2 (1).

Chakma, P.B., (1991), Industrial Workers Commitment in Bangladesh: A Search for Correlates, *Dhaka University Journal of Business Studies,* Vol. 12 (2).

Chowdhury, A. Mannan, (1983), The Settlement of Disputes—A Study of the Jute Industry of Bangladesh, *Journal of Business Administration,* Vol. 14, No. 13.

Cole, G.D.H., (1962), *An Introduction to Trade Unionism,* George Allen & Unwin Ltd., London.

Crouch, Harrold, (1979), *Indian Working Class,* Schin Publications, Ajmer.

D'Costa, Ronald, (1963), *The Role of Trade Union in Developing Countries: A Study on India, Pakistan and Ceylon,* Havaux, Nivelles.

Filippo, Edwin B., (1980), *Personnel Management,* McGraw Hill Book Co., New Delhi.

Fryxell, Gerald E. and Gordon, Michael E., (1998), Work Place Justice and Job Satisfaction as Predicators of Satisfaction with Union and Management, *Academy of Management Journal,* Vol. 32, No. 47.

Giri, V.V., (1962), *Labour Problem in Indian Industry,* Asia Pubshing House, Bombay.

Hussain, Ismail M., Hye, Syed A. and Ali, A.M., (1998), Strutural Adjustment Policies and Labour Market in Bangladesh, in Mustafa K. Mujeri (ed.), *Centre on Integrated Rural Development for Asia and the Pacific* (CIRDAP), Dhaka.

Hussain, M.M. and Haque, M.E., (1991), Socio-Economic Background and Characteristics of Trade Union Leaders at the Basic Union Level, *The Islamic University of Studies,* Vol. 11, No. 1.

Hutt, Allen, (1962), *British Trade Unionism,* Lawrence and Wishart Ltd., London.

Industrial Ordinance, (1969), with commentary, M. Shafi, Nov. 1969, Bureau of Labour Publications, D. Zam Publications, Karachi.

Jha, Shiva Chandra, (1970), *The Indian Trade Union Movement*

Jose, A.V., (2000), The Future of the Labour Movement : Some Observations on Developing Countries, *The Indian Journal of Labour Economics,* Vol. 43, No. 1.

Karmoker, Shyam Sundar, (1992), Factors Influencing Industrial Relations System (IRS) : A Framework for Analysis, *Cost and Management,* March-April, 1992.

Karmoker, Shyam Sundar, (1992), The Quality of Working Life (QWL) in the factories of Bangladesh, *Dhaka University Journal of Business Studies,* Vol. 13 (1).

Karnik, V.B., (1960), *Indian Trade Unions – A Survey,* Labour Education Service, Bombay.

Karnik, V.B., (1982), *Trade Union Movement and Industrial Relations,* Somaiya Publishers Pvt. Ltd., Bombay.

Khan, Abbas Ali, (1962), Patterns and Causes of Industrial Disputes in Bangladesh, *Dhaka University Journal of Business Studies,* Vol. 13 (1).

Khan, Abbas Ali, (1993), Approaches to Resolving Industrial Conflict: A Theoretical Analysis, *Journal of Business Administration*, Vol. 19, No. 1 & 2.

Khan, Abdul Awal, (1990), The Trade Union Leadership and Management in Bangladesh-A Study in Their Relationship, *Chittagong University Studies (Commerce)*, Vol. 6, 1990.

Khan, Md. Abbas Ali, (1986), Government Policies Towards Labour in Bangladesh : A Historical Analysis, *The Dhaka University Studies*, Part C, 7 (2).

Khuda-E-Barkat and Barkat, Abdul (1992), Growth and Structure of Labour Force in Bangladesh, *Dhaka University Journal of Business Studies*, Vol. 13 (1).

Mathur, A.S. and Mathur, J.S., (1962), *Trade Union Movement in India*, Chaitanya Publishing House, Allahabad.

Millen, Bruce H., (1967), *The Political Role of Labour in Developing Countries*, The Brookings Institution, Washington D.C.

Mishra, Lakshmindar (2001), Can Globalisation and Labour Right Co-exists? *The Indian Journal of labour Economics*, Vol. 44, No. 1.

Muniruzzaman, A.K.M. and Shamima, Parveen (1997), Collective Bargaining in Bangladesh, *The Business Review*, Vol. 3, No. 1 & 2.

Nurullah, Sheikh Md., (1985), Patterns of Trade Unions in Bangladesh, *The Rajshahi University Studies*, Part A, XIII.

Nurullah, Sheikh Md., (1965), Industrial Relations System in the Public Sector in Bangladesh: A Case Study of the Textile Industry, Rajshahi University Studies, Part-C, Vol. 3.

Panadhikar, S.G., (1933), *Industrial Labour in India*, Longman Green & Co. Ltd., Bombay.

Rahman, M.A., (1967), Organised Labour and Politics in Pakistan, collected *Seminar Papers on Labour Unions and Political Organisation*, University of London, Institute of Commonwealth Studies, January-March, No. 3.

Rahman, M.A., (1969), Structure of Trade Unions in Pakistan, *The Pakistan Economic Review*, Vol. 1, No. 1, Winter.

Rahman, M.A., (1969), Trade Union Management and Leadership in Pakistan, *Eastern Workers*, Vol. IX, No. 6-7, June-July.

Ramanujam, G., (1986), *Indian Labour Movement*, Sterling Publishers Ltd., New Delhi.

Rayback, Joseph G., (1966), *A History of American Labour,* Coolier McMillan Ltd., London.

Roberts, B.C., (1962), *Trade Union in a Free Society,* Hutchinson and Co. Ltd., London.

Saha, Avinay Chandro, (1986), Research and Development in Jute Industry, *The Bangladesh Accountant,* July-September, Journal of the Institute of Chartered Accountant of Bangladesh.

Saha, Panchanan, (1978), *History of Working Class Movement in Bengal,* Peoples Publishing House, New Delhi.

Saha, Sudhir Kumar, (1993), Managing Human Resources: China vs. the West, *Canadian Journal of Administrative Science,* 10 (2).

Sahoo, Basudeb, (1999), *Problems and Prospects of Labour Movement in India,* Rawat Publications, Jaipur.

Sengupta, Palas and Moniruzzaman, (1997), Industrial Disputes in Bangladesh, Trends, Causes and Impact, *South Asian Journal of Management,* April-June.

Sharma, G.K., (1971), *Labour Movement in India : Its Past and Present,* Sterling Publishers (P) Ltd., New Delhi.

Siddique, Safiq Ahmad, (1989), The Imperial Interest and the Development of the Industrial Relations Systems in Colonial India, *The Dhaka University Studies,* Part C, Vol. 10 (2).

Taher, Md. Abu, (1997), Effectiveness of Conciliation Machinery in Bangladesh, *Journal of Business Administration,* Vol. 23, No. 1 & 2.

Taher, Md. Abu, Moniruzzaman and Mahmood, M.M.H., (1998), Multiplicity of Trade Unions in Bangladesh: An Exploratory Study, *Islamic University Studies,* Part C., Vol. 1, No. 1, June.

Trade Union Section, Department of Labour, Government of Bangladesh, 1999.

Webb, Sidney and Beatrice (1934), *History of Trade Unionism,* Longmans, London.

World Book of Encyclopedia, (1993), Vol. 19, World Book Inc., a Scott Ferzer Company, London.

Chapter - 2

EVOLUTION OF TRADE UNION MOVEMENT IN THE INDIAN SUB-CONTINENT DURING THE BRITISH RULE

Introduction

Labour is the outcome of factory system and trade union is the outcome of the emergence of labour. With the establishment of factory system labour grows as a class, and it happens in case of India too. Factory system had been installed in India only in the middle of 19th century, with the erection of one cotton mill in Bombay and a Jute mill in Calcutta in 1851 and 1854 respectively. Trade union movement in the Indian Subcontinent was late in comparison to western countries, as factory system had been established lately in India. But India had a long reputation for her rich heritage of cottage industry.

Industrial labour had grown in India through an evolutionary process. From coal mining Industry in 1775 to plantation industry in early 19th century, cotton, Jute and other industries came up through gradual process. With the process of industrialisation labour as a class evolved.

In this chapter we will try to discuss the emergence of labour through the process of gradual industrial development of India and Bengal as well as labourers' socio-economic conditions and struggle against the arrogant foreign and native masters. Attempt has been made to depict the growth and development of trade unions and their fight for the legal recognition. They had to fight hard from village panchayat to the formation of all India trade union congress. We will also take up the coming up of labour as a class with gradual industrial development of the country and their socio-economic conditions. The trade union movement of this subcontinent has been discussed in different phases up to 1947.

Emergence of Labour Class

The evolution of labour in British India needs a trace back to the coal mining industry. Production of coal had started as far back as 1775, but insignificant in quantities. It was only the beginning of the second half of the 19th century that production of coal rose to any significant tonnage (Ramanujam, 1986:6). Whatever may be the quantity we can see the existence of coalmines in the last quarter of 18th century in Bengal. So the coal mining industry is older than any other industry in Bengal (Saha, 1978 : 10). Moreover, coal is a labour intensive industry. Therefore, it can be said that invention of colliery gave birth of industrial labour in Bengal and India as well. ILO report of 1838 on industrial labour of India recorded:

> The production of coal is the most important mining industry. The first mine was opened at Raniganj in 1828, but it was not until 1854 when the East India Railway reached the coalfields that the industry began to develop (ILO Report on Industrial Labour in India, 1938 : 24).

The coal production in Burdwan district on an average was about 14,000 to 15,000 tons in 1832 (Dutt, 1956: 286-287). About 95 percent of total coal obtained from Bihar, Orisa and West Bengal in which Raniganj, Jharia, Bokaro, Giridi and Karanpura fields were concentrated. Out of these the first was the most important yielding 77 percent of the country's output. Coal-fields of Central Province, Assam and Beluchistan provinces were unimportant (Panadikar, 1933 : 93).

We has noted earlier that the coal mining industry is situated in Asansol-Raniganj area of Burdwan in Bengal. Dwarakanath Tagore, the grandfather of Rabindranath Tagore had a coal mine at Raniganj colliery. It was the largest coal mine of India at that time. His Excellency pointed out in a letter to the secretary, Government of Bengal, on August 15, 1884 that newly formed Bengal Coal Company expected to bring to Calcutta twenty-lakh mounds coal annually by employing over five thousand persons at the mines and 1500 boats with crew totaling 9,000 men to bring the coal to Calcutta (Quoted in Saha, 1978: 10).

Coal mining did not take place in an organised form before 1895. In 1895 coal production reached three million tons for the first time, whereas since 1906 it had more than doubled (Saha, 1978: 10). The labour force in the coal mines steadily increased since 1890s, with the production. In coal mines labour had to be recruited from the villages around the coal mines. The number of miners was 229,100, who were mostly the natives in 1939, (Sen, 1981: 314) but minorities were obtained from Bihar, United Province and the Central Province. Being more intelligent, adaptable, painstaking and regular than the aboriginal workers, they were being increasingly employed for the coal-cutting machinery, explosives and other modern and scientific methods of mining and of regular working hours (Panadikar, 1933: 95).

The number of collieries under the scope of the Indian Mines Act was 208 and the number of persons employed in them were 44,303 during the year 1929 and the total output of this year reached up to 5,965,104 tons. But wages were at about the same level as in the previous year (Report of Administration of Bengal, 1929-30: 84-85). So, exploitation and deprivation started since the beginning of industrial system came into being. Since the beginning the capitalists aimed at maximizing the level of profits, thus depriving the working class.

The Table 2.1 below gives a picture of coal production in each Indian province in 1881.

The total population of India was 191, 691, 731 in 1901, while the population supported by industry was 45,719,922 and population supported by commerce was 4,197,771 (Census of India, 1901, Vol. 1, Part-1, Report : 224-225) and the number of actual workers in collieries, minerls and other subordinates were 62,485 of which the number of female workers was 27,962 (Census of India, 1901, Vol. 1, Part-1, Report : 231). In 1911 the total population of India was 313,470,014 and population supported by mining was 375,927 respectively. In 1921 it was 316,055,231 and 398,968 (Census of India, 1921, Vol. 1, Part-1, Report : 284). The following table 2.2 renders a clear picture of population supported by industry, coal mining and mining as a whole.

The total quantity of coal exported during the year 1832-33, was 451-564 tons which earned rupees 43,68 lakhs (Report of Admn. of Bengal, 1932: 33-163). Since coal is a labour intensive industry, therefore, it can be said that invention of colliery is the birth of industrial labour in Bengal and India as well.

Plantation industry was one of the oldest industries in India, and indigo came first as plantation industry, especially in Bengal. B.B. Kling says—

> "Since pre-historic times indigo had been grown and processed in India, and during the Roman Empire and the Middle Ages small quantities had been exported to Europe. It was one of the rare tropical products which first attached European traders to India"... (Kling, 1966: 16).

Europeans learned the indigo plantation from India when Alexander (327-324 B.C.) conquered Punjab. It was introduced in Europe from India, and widely used in European textiles especially in England after the adoption of blue for British Naval costumes. However at a latter stage the production considerably improved in West Indies and other South American colonies (Ahmed, 1978:2). It was very interesting that from 17th to 20th centuries indigo was a fugitive among industries. Indigo reached up to the West Indies from India and then back to Bengal in Eastern India. Finally it took refuge in Bihar from Bengal (Kling, 1966: 15-16).

Table 2.1 : Production of Coal in Each Province and State in India (1878-1915)

(In Tons)

Year	*British Provinces*									
	Burma	*Assam*	*Bihar & Orissa*	*Bengal*	*Punjab*	*N.W.F.P.*	*Baluchistan*	*Central Province*	*Madras*	*Total*
1878-1880 (Average)	-	-	-	946,000	-	-	-	41,000	-	987,000
1886-1890 (Average)	-	105,000	+	1,411,000	16,000	-	5,000	137,000	-	1,674,000
1896-1990 (Average)	12,000	201,000	-	3,673,000	83,000	-	18,000	151,000	-	4,228,000
1910	-	297,236	7,041,208	3,737,322	49,189	90	52,614	220,437	-	11,398,096
1915	-	294,484	13,680,030	5,302,295	50,418	210	43,125	481,470	-	19,852,062

+ Figure included in Bengal

Year	*Indian States*					
	Hyderabad	*Rajputana (Bikaner)*	*Central India (Rewah)*	*Kashmir*	*Total*	*Grand Total*
1878-1880						
(Average) 1886-1890	41,000	40,000	101,000	-	298,000	
(Average) 1896-1900	378,000	3,000	141,00	-	522,000	
(Average) 1910	506,173	12,744	130,400	-	649,317	
1915	659,122	11,343	199,975	-	870,431	

Source : Statistics of British India, 1921, Vol. 1, Commercial, p. 88.

Table 2.2 : Population Supported by Industry, Mining and Coal Mining (Male & Female)

Total Population	*Population Supported by Industry*	*Population Supported by Mining*	*Population Involved in Coal Mining*	
			Male	Female
316,055,231	33,709,071	375,927	131,247	74,024

Source : Census of India, 1921, Part-I, Report, pp. 282-285.

A different information we get regarding plantation of indigo in Bengal. At the end of 18th century the industry was imported into India from West Indies by European planters. One French trader Louis Bonnard introduced indigo cultivation in India in 1777. Among the English traders Carel Blume first introduced the indigo cultivation in Bengal in 1778 (Bose, 1969: 216). B.B. Kling also describes almost in the same manner—when the West Indies planters voluntarily switched over to the more profitable commodities of sugar and coffee, the British coal manufacturer were forced to depend on the supply from Spanish Guatemala, French Santo Dominigo and Southern Carolina. These sources of dye were lost in the 1770s, with the American Revolution, and the British manufacturers were forced to pay more for indigo. The directors of East India Company therefore undertook steps to promote the cultivation of these cash crops in Bengal (Kling, 1966 : 16-17). Afterwards Company discovered that indigo produced in Bengal under European supervision was equal in quality to the finest West Indian product. Then the East India Company helped the planters with advances. The Company also encouraged experienced West Indian planters in Bengal. The governor General John Shore carried out a highly effective measure to improve the quality of indigo exported from Calcutta and stimulate the industry in Bengal in 1796. He imposed 15 percent duty on all indigo brought into the province from Agra, and Oudh, areas which until that time supplied one fifths indigo exported from India to London. At the end of the century the amount of indigo exported to U.K. from Bengal was more than double of the quantity from all other sources (Kling, 1966: 18). Romesh Dutta said more categorically "the importation of indigo from India commenced about 1790 and had so greatly increased in forty years as to supersede all other indigos" (Dutt, 1902: 281).

The scenario of cultivation of indigo was different than that of tea or other plantation in Bengal. Two types of indigo cultivation, *nijabad* and *raiyati* was in practice in lower Bengal. In *nijabad,* as in a conventional plantation system, that factory supplied the land and seed and hired the cultivators, usually along with ploughs and bullocks. *Nijabad* cultivation took place only on the adjacent lands of the factory and on *chars* (muddly plains formed due to changing course of the rivers). The major indigo cultivation was under *raiyati* cultivation outside the factory grounds. *Raiyati* land might have belonged to Indian Zamindar (*be-ilaka*) or the planter himself or it might be the Zamindar (*ilaka*). When it was cultivated on *raiyati* land that indigo conflicted *raiyats* and Zamindar (Kling, 1966: 29).

In northern and eastern Bengal where *char* lands were more plentiful, there was higher proportion of *nijabad,* but in lower Bengal as a whole almost three times as much land was cultivated under *raiyati* as under *nijabad.* Even they used to force the small and medium landholders to take the advances for the cultivation of indigo. The system by which European indigo manufacturers gave advances to the *raiyats* for cultivation of indigo was bound to lead to fraud and oppression (Sinha, 1961: 228). Advances were made to the tenants of local landlords to cultivate indigo and once they took advance they were no longer freemen and doomed forever. So cultivators were very much reluctant to produce indigo under planters' terms and conditions which were never profitable but they were forced to do it on slashes of whip. On the contrary it was very much profitable to the planters. As it was profitable there were nearly one thousand Europeans in the 400 and odd indigo factories in Bengal in 1831 (Ahmad, 1978:2). But due to this dis-satisfaction of cultivators Blue Mutiny took place in Bengal and continued from 1858-60. The agitation was particularly widespread in Jessore, Nadia, Pabna, Faridpur, etc. (Bose, 1969 : 217). Due to this indigo rebellion we could follow a marked decline in the cultivation of the plant throughout the Bengal. There were 300 or 400 factories in Bengal chiefly in Jessore, Krishnagar and Tirhoot. All the indigo factories were destroyed during that rebellion (Dutt, 1956: 280). So the roaring business declined miserably. The industry recovered the decaying condition in Bengal due to its extension in Bihar and the North West Provinces (Sinha, 1961: 355-256).

Actually there were no regular labour employed in the plantation of indigo but in the indigo factories some skilled workers, employees and very small amount of labour were required. During the manufacturing season *Buna* coolies were brought from jungle tribes of Manbhum, Singhabhum and Midnapore and were employed. A large factory like Mulnath, head quarters of Bengal indigo company, employed one hundred of them in each manufacturing season paying 3 or 4 Rupees per month and local labour was also employed to operate the pumps, boilers, cutting machines and other equipments as well as cart men and boat men were hired to bring the indigo plant to the factory (Quoted in Kling, 1966: 29).

So we can say that the planters used to cultivate indigo with the help of the tenants of local landholders under pressure from different corners. Neither labours nor tenants used to work willingly in the indigo factories or in the field of the foreign masters or their close allies, Zamindars of Bengal.

Some plantation industry, which has less onerous system of indentured labour, existed at that time. One of the largest plantation industries of Indian subcontinent was tea. Its development necessitated two basic requisites-firstly, large areas of cultivable land and secondly, huge labour force. Moreover the areas suited for plantations were initially sparsely populated, so plantation had to depend on migrated labour all through like other plantation areas of the world (Bhowmik, 1981 : 10-11). Tea plantation requires huge labour, as it is a labour intensive and agro industrial product.

The growth of the tea plantation of India is very interesting. Major Robert Bruce discovered indigenous tea plants in Assam in 1823. During the following decade discovery of wild tea was made in Assam and in 1834 Lord William Bentinck appointed a commission charged with the duty of submitting to Government "a plan for the accomplishment of the introduction of tea cultivation in India and for the superintendents of its execution (Quoted in Saha, 1978: 7). On the basis of the report an experiment was made on the lower hills and valleys of the Himalayan range and the experiments were proved successful and it was found to be highly profitable (Quoted in Saha, 1978 : 7).

The experimentation in cultivation of tea in India was undertaken by the Government in 1834. Tea makers and artisans from China were introduced in 1837 and some consignments of manufacture of Assam tea were forwarded to the court of Directors in the year 1838-39. To see the high quality the English Mercantile world and the Company were attracted for the cultivation of the tea plant and manufacture of tea in upper Assam. The first trial of tea plant at Darjeeling was made in 1841. And these ventures also came out successful (Gosh, 1978: 9). The discovery of indigenous tea in Sylhet and Cacher gave the impetuous for an expansion of the industry into Surma valley, and few years thereafter the whole of the upper portion of the province of Assam (both Brahmaputra and Surma valley) was converted into large tea plantation (Saha, 1978: 8-9). The tea plantations were localized mainly in Assam, Bengal and also in parts of U.P., Punjab, and in some southern states (Gosh, 1978: 3).

Now the question arises why the tea plantation industry was encouraged by the British. Firstly, there was a huge demand for tea in the Indian market and thus a big amount of it had to be imported from China. Secondly, tea had a demand in the world market, especially in the western world. Thirdly, Warren Hestings realized that the growth of tea industry would promote industrial development of India (Bhopwmik, 1981:7) and lastly cheap labour, land and favourable climate for tea production, and foreign capital ushered in the growth of tea plantation.

In May 1838 the first consignment of eight chests of tea from India was supplied to London and was sold by auction at prices ranging from 16 Shilling to 34 Shilling per pound. Another important news was that the Assam Tea Company was founded in 1839, and it was the first tea company (Saha, 1978: 8).

How tea plantation industry came up is not our concern, rather the emergence of working class through the gradual flourishment of industrial sector is our main issue of discussion. The work force of tea plantation industry is a bit different from any other industry. Tea gardens were generally located in jungle lands where there were wild animals, mosquitoes and other poisonous insects too. So working conditions at the growing stage of the garden were very hazardous and risky. Local people having a minimum means to live were disinclined to work in the tea gardens. In this situation all the tea producing areas had to depend on immigrant workers of Bihar, Orrisa, U.P., Santals from Santal Pargana, Nepal, and some local labour. Not only in the tea plantation of India, plantation industry all over the world had to depend on migrant labour like cotton plantation in North America, rubber in Malaysia and tea in India (Bhowmik, 1981: 11).

An exquisite report on 'Industrial Labour in India' conducted by International Labour Office of Geneva, presented valuable information regarding plantation of tea in Assam and Bengal as follows:

> "The plantations in the North area principally located in Assam and Bengal. In Assam the plantation are distributed in the Brahmaputra or the Assam valley in the North and Surma valley in the South. The planting areas in Bengal comprise Darjeeling, Terai and the Dooars in the North and Tripura in the South-east. In Darjeeling the tea gardens occupy slopes from 6,000 feet downwards; Terai is level country at the base of and surrounded by high mountains and is only about 300 feet above the sea level; the Dooars is a sub mountain country, South of Bhutan and 22

miles in width, between Tista and Sankosh River. With the exception of Cinchona, which is grown by the Government in the district of Darjeeling, the only plantation crop is tea. A few small plantations are also to be founded in Punjab, the United Province and Bihar and Orissa" (ILO Report, 1938: 36).

Regarding the manufacture of tea we can say that there was steady progress marked in this industry. The quantity of tea manufactured was 111,355,903 Lbs in 1928. The number of tea plantations totaled 388 and employing on an average 194,511 permanent and 9,432 temporary hands daily (Report of Administration of Bengal, 1929-30: 84).

Tea was ranked second to jute in the number of persons employed among the organised industries of Bengal. Virtually tea-producing districts were Darjeeling and Jalpaiguri in Bengal. Since tea plantation started in Bengal in 1850s its growth had been phenomenal. The industry was the first in Bengal as a commercial enterprise in Darjeeling in 1856. In 1870 there were only 56 gardens, employing about 8,000 labourers in Darjeeling and by 1901 there were 148 tea gardens, employing 64,000 coolies (Bengal District Gazetteers, Calcutta, 1907: 44-45).

From the 1911 census it is evident that the work on tea gardens appealed to a very large number of tribal not less than 91, the chief among them being Oraons, Mundas, Jeondars or Khambus, Santals, Murmis, Mangars and Kharias. The nature of labour force varied in different gardens (Broughton, 1924 : 77). It varied because the sources of labour in different plantation areas were different.

Local Assamese people were very much reluctant to work in the tea gardens. This necessitated the employment of labour from outsiders, like, tribes, people belonging to the Hindu untouchable community from the district of Ranchi and its neighbourhood and from eastern part of U.P. and Madhya Pradesh with false hopes and aspiration. For recruitment of labour from distant parts of the country for Assam gardens passed two special Acts, first was the workers breach of contact Act, 1859. Under this Act planters could take summary remedy against deserters. The second was the plantation Act 1863 requiring the recruiters to be licensed, the emigrant to be registered, and sanitation could be provided on the way to labour districts. Desertion on the part of the workmen was made possible and planters were empowered to arrest absconders without warrants (Ahmad, 1978 : 3). So the conditions of workers were like that of slaves. And these labourers had to work hard in jungles with minimum wages and little fringe benefits and above all tuckered in different ways. Not only in Bengal, in India vast majority of workers got a wage which was not enough to provide them with the meanest necessities of life (Dutt, 1969 : 388).

From the following table (table 2.3) we would get the number of labour employed in specified years of Assam tea gardens.

Table 2.3 : Labour Population Living in Assam Tea Gardens in Specified Years

Year	*Adults*	*Children's*	*Total*
1877...	110,675	46,544	157,219
1900...	410,075	252,374	622,451
1919-20	629,631	480,733	1,110,364
1935-36	618,043	521,353	1,139,396

* The year chosen are those when the number of workers on the gardens was large.

Source : ILO Report on Industrial Labour in India, 1938, p. 38.

From the following table (table 2.4) we can have and idea about the growth of tea plantation workers of both Bengal and Assam.

Table 2.4 : Growth of Tea Plantation Workers

Place	*1935*	*1942*
Assam	519,621	521,667
Bengal	205,428	221,251

Source : Saha, 1978 : 9.

From the following table 2.5, we can get a picture of number of tea gardens employing more than 20 persons.

Table 2.5 : Gardens Employing More Than 20 Persons

Year	*No.*	*Persons*
1911	1,002	703,585
1921	1,353	746,760

Source : Census of India, 1921 : 266.

In Jalpaiguri tea gardens, the majority workers were found to be immigrants, more than half coming from Chota Nagpur (Census of India, 1911: 358). The growth of Jalpaiguri garden is significant as it started with 13 gardens in 1876, and rose to 235 in 1901, employing some 90,000 persons (Bengal District Gazetteers, Jalpaiguri, Allahabad, 1911: 102-103).

In the gardens of Darjeeling almost half of the labour was local, the bulk of the remainder coming from Nepal (Broughton, 1924: 77). It is the special characteristics of Darjeeling tea gardens that there were some local labours.

The valuable information regarding labour and production in the tea gardens is revealed in the Statesman's year book of 1918 as follows:

One of the most important industries connected with agriculture is tea industry, the number of persons employed being about 698,000 in 1915-16. The area under tea plucked in this year was about 585,000 acres, distributed as follows: Assam, 359,000; Bengal, 151,000; Madras, 21,000; Punjab, 10,000; Agra, 8,000; Bihar and Orissa, 2,000; Upper Burma, 1,600; and the Travencore State, 33,000. The production of 1915-16 was about 372 million pounds (Statesman's Year Book, 1918: 136).

Regarding the daily working people in tea factories a notable statistics we can get from a report on the administration of Bengal 1932-33:

> There were 288 tea factories under the Tea Factories Act, and the total daily average number of operatives employed was 15,722.

This number excluded those employed in gardens outside the factory (Report of the Admn. of Bengal, 1932-33: 143).

In tea plantations both male and female, child and adolescents are working. Employment of adolescent and child labour is a special feature of the plantation industry especially in Assam and West Bengal, but it is minimum in South India. The tea industry in general has more female workers than male workers (Bhowmik, 1981: 5). And this trend is followed in most Asian countries. In the then Ceylon and Vietnam woman account for over 50 percent of the labour force in plantations. In Malay and India the Share is over 40 percent and in Pakistan and Philippines it is around 35 percent (Boserup, 1970: 76). The table 2.6 presents the tea producing areas and number of workers area-wise with age group and sex.

Table 2.6 : Employment of Workers in Tea Garden in 1973

Place	*Male*	*Female*	*Adolescent*	*Children*
Dooars	61,358	57,959	5,335	8,040
W. Bengal	85,032	92,901	6,668	11,558
N. India	275,314	277,354	19,172	41,474
S. India	55,461	81,924	3,052	1,285
All India	341,775	359,278	22,242	42,759

Source : Bhowmik, 1981 : 6.

The exports of Indian tea from British India (including the State of Travancore) in 1915-16 were: to United Kingdom, 250,290,000 lb.; China, 9,865,000 lb.; Australia, 9,591,000 lb.; Ceylon, 4,307,000 lb.; Asiatic Turkey, 2,327,000 lb.; United States, 3,443,000 lb.; elsewhere (including exports across the land frontiers), 15,178,000 lb.; total 340,433,000 lb. (Stateman's Year-Book, 1918 : 136).

Though coffee was not a common beverage like tea, but it was one of the important plantation industries of South India. Coffee was imported from Arabia and Abyssinia (through Mocha) as a costly beverage taken by high society in Mughal period. But unlike tobacco it did not become popular before the second half of the 17th century. (Sarkar, 1987: 28).

Coffee was cultivated extensively only since 1823. The Government encouraged coffee cultivation by permitting the planters to hold land for a long series of years. Coffee was produced in Bengal also. In Bengal 400 acres had been led out in coffee. Here the sun was too powerful for coffee cultivation (Dutt, 1956: 283).

As the tea plantation is the most important industry in Eastern India, coffee plantation became the centre of activities in South India (Chowdhury R.C., 1987 : 88). The most important areas for the coffee industry are the Madras Presidency and Coorg in British India and the States of Mysore, Travancore and Cochin; in 1935-36, the number workers was 31,655 in the Madras Presidency, 21,623 in Coorg, and 48,117 in Mysore (ILO Report on Indian Labour, 1938 : 37).

Most of the coffee was exported to European countries like France and Norway. It may be mentioned that whereas tea is consumed mostly in U.K. and her dominions and coffee in European countries other than U.K. and America. The exports of coffee were earlier in origin than those of tea from India, and had greater importance for sometime, but with the growth of tea industry, coffee got a lower position (Vakil *et al.*, 1931 : 57). Principally, the whole of the production was exported. The following table 2.7 shows the figure of the trade during five years.

Table 2.7 : Figure of the Trade of Coffee in Five Years

Year	*CWT*	*Rupees*
1899-1900	281,353	1,48,47,146
1900-1901	264,431	1,22,84,498
1901-1902	255,042	1,25,02,200
1902-1903	269,165	1,32,12,628
1903-1904	291-254	1,36,73,773

Source : Review of the Trade of India, 1903-04.

Through these economic activities labour force was emerging gradually in Bengal and India as well. People were getting employment with the development of agro-based industries side by side with the agricultural sector.

Jute was regarded as the golden fiber of Bengal, and it was growing in Bengal for a long time past, especially in East Bengal. Bengal was the largest producer of jute in the world. But Bengal jute was mostly used in the early 19th century. Later on the use of jute began to increase in place of flak and hemp. Thomad Neish, a Dundee merchant, persuaded the Dundee spinners, Messers Bell and Balfour, to try spinning jute with their machinery. The experiment was successful. 18 tons of jute was imported by the Dundee manufacturers in 1828. The figure rose to three hundred tons in 1933. By then a large part of flax machinery had been adjusted to jute, and in 1833 saw the beginning of the mechanical manufacture of jute yarns and fabrics (quoted in Chakrabarty, 1996: 16). Pure jute yarn was put up for sale for the first time in 1835 and within a few years jute appeared in the European market as a competitor of flax and hemp (Ahmad, 18966: 16). Raw jute was exported in the early days of East India Company to Dundee where jute goods manufacturing industry had been developed rapidly. But later on company thought that labour was very cheap, so they could start mills in Bengal which would earn more profit.

Use of jute increased appreciably by the decision of the Netherlands Government by about 1833, to replace flax by jute in the manufacturing of coffee bags for East Indies. And the Crimean war (1854-55) cut off the supply of flax, the main source of which was Russia. So the Bengal jute began to flow in European market and it became popular (Sinha, 1967: 255).

Mr. George Auckland, formally a coffee planter of Ceylon started the jute-spinning mill – the Rishra Yarn mill component at Rishra near Srirampore in the Hoogly district in 1854. It started with the production capacity of 8 tons of yarn per day. Two years later the first weaving mill was established at Baranagar. The name of the mill was Borneo Company which, came later on to be known as the Baranagar jute factory limited. This company first installed 192 power looms in India (Ahmad, *Year Book on Jute,* 1967: 17-18). With the establishment of first jute mill in 1854 in Bengal and first cotton mill in 1851 in Bombay, factory system had been introduced in Bengal as well as in India. This was the real birth of industrial labour in India in the truest sense. With the establishment of factory system, industrial labour force began to grow slowly.

Jute mills began to increase gradually as it was very much profitable. The five jute mills were established between 1855 and 1866. And by 1872-75 thirteen new mills were set up and the number of looms increased from 1,250 in 1873 to 3,500 in 1875 (Sinha, 1967: 255). Almost all the jute mills were set up around Calcutta. From 1879-80 to 1935-36, the number of mills rose from 22 to 104, and the number of looms and spindles increased from 4,946 and 70,840 to 63,700 and 1.27 million respectively *(Statistics of British India,* Vol. 1, 1922 : 76).

With the establishment of first jute mill in 1854 setting up of new mills was marked, as it was a profitable industry. The data presented in table 2.7 shows the progress of the jute mills.

Out of a total of 243,000 tons of jute exported during the financial year 1942-43, 89,000 tons were consigned to the United Kingdom and 124,000 tons to the United States of America. Brazil, U.S.S.R., Argentina, Canada and Australia were other important customers for Indian raw jute *(The Review of the Trade of India in 1942-43: 15).* Table 2.8 shows the export of raw jute and manufactured jute goods from India.

The number of workers in jute and other fibers, with their families, had risen from 461 to 649 thousand. About one third of the total area operatives, etc, in the Bengal jute mills and jute posses, who aggregate 143,000 compared to only 38,000 ten years ago (Census of India, 1901: 214). Population supported by jute hemp, Flax, Coir, etc. were 461,193 and 649,604 in 1891 and 1901 respectively (Census of India, 1901, Vol.1, Part-I, Report: 239).

Table 2.8 : Quinquennial Progress of the Jute Mill Industry in India from 1879 to 1915 and 1918-19

Year	*Number of Mills at Work*	*Persons employed*
1879-1880	22	27,494
1884-1885	24	51,902
1885-1890	26	59,541
1890-1895	28	74,357
1895-1900	34	102,440
1900-1905	38	133,162
1905-1910	60	204,104
1910-1915	70	238,274
1918-1919	76	275,500

Source : Statistics of British India Vol. 1, Commercial Statistics, 1921, p. 74.

Regarding the labour force in the jute mills Forely reported in 1906 :

> "Twenty years ago all the hands were Bengalies" (Quoted in Sen, Sunil Kumar, 1981: 312).

The Bengali workers who came from peasantry and weaving formed 53.6 percent of the labourers in the jute mills in 1895. The number of Bengali workers began to decrease when a good number of labours started coming from Bihar, United Province, Orisa and joined the industry (Sen, 1981 : 312). There might be a question why there was a continuing flow of immigrant labour in Bengal? There were so many reasons for this influx. Bengal is a fertile land with green vegetation, moderate weather and easy going life than other regions of India. In British period Calcutta was the capital till 1911 and Bengal Presidency was the most important presidency than Madras and Bombay. Political, economic and cultural activity of India was controlled by Calcutta. So people all over India used to come to Bengal in search of job, who settled permanently. Moreover, the recruited contractors and sardars were non-Bengalees. They were involved in regional nepotism, encouraging the concentration of migrants from the same area in the same industrial centers (Sen, 1981 : 313).

Table 2.9 : The export of Jute & Jute Goods in Three Consecutive Years (In thousand of Rupees)

Items	*1940-41*	*1941-42*	*1942-43*
Raw Jute	7,84,60	10,41,78	9,01,57
Jute manufactures	45,38,46	53,89,54	36,40,93
Total Value of export	53,23,06	64,31,32	45,42,50

Source : Review of the Trade of India in 1942-43, p. 80.

Classification of jute mill employees according to castes was not vivid in the 1921 census. Only 21 castes were stated while at the census of 1911 no less than 71 castes were enumerated. According to the censuses of 1911 and 1921, Muhamadans constitute a substantial proportion of the workers. Muslims were called *Jolahas* and Hindu weavers were known as *Tanti* and *Tatwa* (Broughton, 1924: 77).

The data shown in table 2.9 above will give a clear picture of jute workers in the province of Bengal according to caste or race.

The Baranagar Jute Factory Limited installed the first 192 power looms in India. In 1946-47 the number rose from 1 to 108 and that of looms from 192 to 11,300 (Ahmad, 1969: 17-18).

The long heritage of highly skilled craftsmanship of Bengal weavers and comparatively low price of produce made its textiles attractive to Asian merchants for centuries (Chowdhury, 1997: 257-58). Not only Asian countries Bengal's cheap and high quality textiles attracted the European countries too. But our craftsmen and artisans were always deprived of proper wage and benefits and services programmes. So they had no general standard of living. They were so ill paid that they had no social status. In order to maintain minimum standard moderate wage was essential.

The handicrafts manufacturing industry of Bengal, especially in the field of textile and silk production was at its peak during the rule of the Bengal Nawabs, supplying the enormous demand of both Asia and Europe (Chowdhury, 1997:

Table 2.10 : Caste or Race of Jute Workers in the Province of Bengal

Caste, Tribe or Race	*Skilled*	*Unskilled*
1. Muhamadan	46,917	33,179
2. Chamar and Muchi	6,941	21,883
3. Tanti	4,811	6,812
4. Chasi Kaibartta	4,022	2,219
5. Teli	3,677	6,143
6. Bagdi	3,546	4,429
7. Jalia Kaibartta	3,489	3,978
8. Goala	2,739	4,454
9. Brahmman	2,154	4,614
10. Unspecified	83,604	58,586

Source : Broughton, 1924 : 76.

31). Not only the Bengal under Nawabs, rather throughout the medieval period this country was famous for its textiles. Robert Orme found the cotton manufacture to be a national industry of Bengal about mid 18th century as there was hardly a village near large towns and on the principal roads where every inhabitant man, woman or child, was not engaged in manufacturing. But this distinction had not grown in a day; it had a tradition of at least several centuries. Besides Tanda, Maldah and Salgaon, Sonargoan (The old Muslim capital of Eastern Bengal) was noted for the production of 'a species of muslin very fine and great quality' in the time of Akbar (Sarkar, 1987: 49).

The export trades of the European Companies Bengal textiles were the most important single manufactured article both in terms of value and volume. Bengal provided more textiles to the European Companies than the rest of Asia put together (Chowdhury, 1997 : 157).

N.K. Sinha added the following–

> "The export of Bengal cotton textiles was major part of the East India Company's commercial activities since it had started its operations in Bengal. It had its manufactories at Santipuram, Dacca, Tandha and Birbhum" (Sinha, 1967: 350).

In the medieval period, Bengal was not only famous for cotton textiles but also silk product in the European and Asian Market. As far as the actual exports of Textiles, were concerned. Dutch company used to export more than the English, British went ahead of the Dutch in early 18th century. During this decade the Dutch and British exported about 203,353 and 278, 588 pieces respectively. Before 1757 the export of Bengal textiles by Asian merchants were larger than those by Europeans. The total export from Dhaka in 1747 the Asian share including that of the Americans stood by two third of the total compared to the one third of Europeans including the European Private trade in 1774 (Sushil, 1997: 260-61). The total value of annual textile export formed roughly about 70-90 percent of the total value of the company's export (1975: 205).

As far as the employment in the handloom cotton textile industry is concerned in 1776 the weaving of the cloth in the Dacca district alone provided employment to some 146,751 individuals which included 25,200 weavers and 8,000 spinners. In and around 1825 there were about 1.5 million in the province who were directly engaged in the production of cotton textiles (Awal, 1997: 355-56).

Production and trade to cotton textile was very much satisfactory, but that was the age of cottage industry; modern factory had not been introduced at that time. The production was carried out with the help of family members or few hired labour in separate rooms and corridors. Weaver was the chief of the family and the production process, and the family members were the personnel of his production unit.

Regarding the establishment of cotton mills in India the Indian Cotton Industry, 1940 annual (The Indian Cotton Textile Industry, 1940: Annual, September, 1940 : 1) says that—

> It was as early as 1818 that the first cotton mill was established at fort Gloster near Calcutta, but the real development of the cotton mills one the modern joint stocks principle in this country dates from 1851 when the Bombay Weaving and Spinning mill was established in Bombay.

But within few decades there are good numbers of mills in India. Regarding the development of the industry this report further notes that:

> ....Towards the end of the last century, we had no less than 156 mills and by 1914 the number had increased to 239 with a total nominal capital of Rs. 20,00, 14,000 and equipped with a total compliment of 90,268 looms and 6,208,756 spindles. By the end of 1939 the total number of equipped mills in this country stood at 389.

This is the beginning of the factory system in India. After that, mills and factories began its establishment in different places of India. By the end of 19th century 144 cotton mills and cotton twines of over 77 million lbs were exported (The Indian Cotton Textile Industry, 1940: Annul, September). Steady progress was made by the textile industry at the beginning of the 20th century as anti-British movement got impetus due to *Swadesi* (self Government) movement and the partition of Bengal in 1905 when it was resolved not to use any cloth made in Lancashire or Manchester cotton textile mills. These movements had a direct impact on cotton textile industries of India.

By 1914 India had become the fourth greatest cotton manufacturing country in the world (Anstey, 1952 : 262). The number of workers working in cotton sectors is shown below in the table 2.11. According to 1901 census report of India progress of cotton mills was satisfactory. We get a more clear picture of the progress from table 2.12.

But as a whole the conditions of labour in Bengal and India remained deplorable. They had to work for 11 or 12 hours a day with half an hour's recess, sometimes in extended up to 14 or 15 hours or more. No extra payment for that—no facilities or amenities of life that a human being deserves. In Calcutta the working time was a bit different in different mills. The position of workers was better than other states. The report of the Indian factory labour commission of 1908 (Indian Factory Labour Commission of 1908: 6-10) describes regarding the working hours[1] in case of textile or other factories in detail.

Table 2.11 : Workers of Cotton Sector According to Axe & Their Percentage

Order of suborder	*Number of actual worker*		*Number of females per 1000 male*
Cotton	*Male*	*Female*	
1. Cotton cleaners, pressers and ginners	153,801	89,942	585
2. Cotton weavers, hand industry	1,386,434	832,594	458
3. Cotton spinners, seizers and yarn beaters	90,295	509,200	5,639
4. Cotton yearn and thread sellers	23,284	10,472	450
5. Cotton dyers	72,854	27,366	376

Source : Census of India, 1901, Vol. 1, Part-1, Report, p. 232.

Table 2.12 : Progress of Cotton Mills in India from 1932 to 1938 (Figure in thousands)

Year ending 31st August	*Number of mills*	*Average No. of hands employed*
1932	339	4,03
1933	344	4,00
1934	352	3,85
1935	365	4,15
1936	379	4,18
1937 (a)	370	4,17
1938 (a)	380	4,38
1939 (a)	389	4,41

(a) Include Burma.

Source : The Indian Cotton Textile Industry, 1940, Annual : x.

Next to textile, raw silk was one of the most important exportable items of the European Companies from Bengal almost in the seventies of 17th century. Indian silk had a long tradition for its high quality and cheap price. Silk was produced in different places in India such as–Malda, Rangpur, Dacca and

Kumarkhali (Nadia) and the Southern India. European Companies especially the Dutch and English were interested in Chinese and Persian silk. In the middle of 1670 they turned their attention to the Bengal silk, which was of high quality and less costly than Chinese, Persian, Italian or Spanish silk. So it was easier to earn huge profit by exporting Bengal silk to English and other European markets. The average annual export of Bengal raw silk to England during the decade 1776-1785 rose to 560,283 (small) lb., while that from Italy, China and other countries did not altogether exceed 282,306 lb. (Sinha, 1927: 175). The major silk production and trade centre in Bengal was Kasim Bazar and its neighborhood including Murshidabad, Bauleah (Rajshahi), Kumarkhali, Malda, Radanagore, Rangpur and Rangamati (Chowdhury, 1997: 249-51).

European companies exported quite a large amount of raw silk from Bengal, but they could hardly control the silk market of Bengal. The large number of Asian merchants were engaged in the silk market of Bengal. European Companies had to face tough competition from the several of Asian merchants from Gujarat, Lahore, Multan, Benaras Gorakhpur, Hyderabad, Delhi ('Calwars') and Jangipur (Murshidabad) and Armenians too. The Asian merchants exported larger amounts of raw silk from Bengal than the total exports of the European Companies taken together in the 1740s and 1750s (Chowdhury, 1997: 252-255).

In addition to cheap labour, Bengali artisans' ability and attitude, finest quality of silk and its colour made Bengal's silk industry famous in Asian and European market. In 1853 a Dutch factory was established in Kasim Bazar for reeling raw silk. When the factory operated in full capacity it would reel about 1,500 bales of raw silk per annum and employ over 300 men. But when the factory faced dull time workers strength came down to less than hundred (Chowdhury, 1997 : 252).

Salt is another essential item widely produced in Bengal. It was manufactured in the costal region of Bengal Presidency by a class of people know as *malangis*[2] whose profession was, by and large hereditary. Salt was produced in two ways—first, by solar evaporation; secondly, by boiling the brine. The second one is expensive and a laborious process. But it was widely practiced in Bengal. People of Bengal mostly used this type of salt. Labourers had to be hired for the whole salt season, in some places, where the labours were scarce; the manufacturers were obliged to pay the entire season's wage (Serajuddin, 1973: 55).

Salt was produced in Bengal along the costal belt between Balasore and Chittagong, in Hijli, Tamluk, the Twenty Four Parganas, Dacca, Noakhali and Chittagong. Bengal produced 30,28,342 mds., in 1795 and 26,99,286 mds. of salt in 1796. On the average Bengal produced about 28 lakhs mds. of salt every year (Sinha, 1965: 216).

There were two types of salt worker (*malangis*) in Bengal at that time namely *ajoorah* and *thika*. The total number of workers were about 60,000 of which *thika malangis* constituted one-third of the total. The *ajoorah malangis* were tied to the soil and there was no recession from it. The house and lands of the *ajoorah* were close to the *Khalari* (salt workers) lands. The manufacture of salt was a family concern. The price of *ajoorah* salt was gradually increased from Rs. 14 per hundred mounds to Rs. 26-10-12 in Hijli and Rs. 27-2-9 in Tamluk. The lowest rate of payment to *thika malangis* was 40 Arcot rupees per hundred mounds in Hijli. Their daily wages amounted to 1 anna 2 gandas and 3 cowries. The common labour got 13 gandas per day (Sinha, 1965: 18-19).

Except all these agro-based manufacturing industries and many other non-agriculture industries there were so many industries in existence in pre-colonial India. The rich heritage of handicrafts or cottage industries, constituted a characteristic feature of rural economy of India and Bengal as well, one of her richest province. We are going to mention some other important industrial products which had national and international importance.

As regards sugar industry, according to historian Abul Fazal, five types of sugar were produced from sugar cane and palm trees by peasants of that time:

(a) Jaggery (Gur); (b) red or brown sugar; refined form of gur, (c) white sugar; powdered fine grained, more refined modern form, (d) white candy, (e) nabat; refined sugar balls. White sugar, chiefly manufactured in Bengal, Orissa, Agra-Bayana, Multan, Ahmedabad and Cambay. According to Abul Fazal, Bengal sugar was ranked first in production. It was taken by river to the imperial capitals, where it would cost 128 *dams* a mound. And it was sent by sea to Malabar. According to French Bernier (1656-58) Bengal sugar in large quantity was supplied to Golkanda and Karnatik, to Arabia and Mesopotamia through town of Mecca and Basra and even to Persia by Bandar Abbas. In the 17th century the production of sugar in Bengal possibly increased with the establishment of new markets for it in Europe and Persia by the Dutch and the English (Sarkar, 1987 : 26-27).

Saltpetre is an essential element of gun powder. The Europeans were attracted to this trade quite early. But in international trade in it as an ingredient of gunpowder originated in the military needs of Europe during the 17th century. In the 16th and 17th century saltpetre was used internally in India for cooling water and as an ingredient of gunpowder, in preparing explosives and fireworks. It was highly prized by the Portuguese, the Dutch, the English and the French for export to their respective countries. Regarding the export of saltpetre, East India Company had the monopoly. During the twelve years of peace from 1763 to 1772, the Company sold on an average 22,620 bags of Saltpetre per annum. The Company's export of Saltpetre to England brought a profit of £5,381 in 1776 (Sinha, 1927: 176-177). Saltpetre was found in marked degree in different districts of Indo-gangetic plains, especially in Bihar. It gave employment to peasants of different places, e.g., Bihar, Bengal and Gujarat (Sarkar, 1987: 69).

Water way is the oldest media of transports. People used to carry necessaries and passengers from one place to another through rafts and country boats from the pre-historic ages.

Making of boats developed gradually through the evolution of civilization. So, shipbuilding industry came up in due course.

The Mughal Emperor Akbar founded and maintained an imperial naval establishment in Bengal. The department was known as *nawara.* A large shipyard was established in Dhaka for building new ships and repairing the old ones during his viceroyalty, Shaista Khan. He also ordered as many boats as was possible to build at the ports of Hugli, Balasore, Murang, Chilmory, Jessore and Caribari (Awal, 1997: 397). Local made ships were far better than British or any other Western ships. It was possible as the materials were cheap and plenty, moreover there were local artifices and labourers in good number and cheap. They may not be considered as marine engineer but shipwrights. They proved themselves to be the excellent ship carpenter.

This industry had created employment for both skilled and unskilled labour. Expanding domestic and foreign trade gave a definite boost to ship building industry. In British period Khidirpur (Calcutta) dock and marine yard had been established in 1780 (Awal, 1997: 398). As the industry flourished number of labour increased side by side. The census figures for 1921, indicate that the total number of persons supported by industry was 33,709,709 in India (Census of India 1921: 282). Rajani Palm Dutt expressed his dissatisfaction regarding the number of working class in India. He says—

> The industrial working class in India, in the modern sense, in not numerically large in relation to the population; but it is concentrated in the decisive centers, and is the most coherent, advanced, resolute and basically revolutionary section of the population (Dutt, 1969: 383).

Other than the above-discussed industries there were numerous types of industries and commercial activities in all over India and Bengal, such as food grains, tobacco, opium, which were commercial crops. Many agro-based industries developed with these products. Some non-agricultural industries like hides, skins and leather goods, wool, house hold utensils, paper, tailoring, pottery etc. made this country economically self-sufficient which enjoyed favorable balance of payment altogether.

Before the advent of factory system labour was scattered all over the country with the small unit of cottage industry. With the gradual industrial development in India labour as a class emerged. But the labours' financial position was at stake. They had to sell their labour at a very low wage. Moreover, in July 1925 the Mill-owners Association announced eleven percent reduction of wages. This decision caused serious strikes that India has yet to witness. Afterwards the wage-cut was withdrawn (Anstey, 1952: 263).

It generally happens that production may increase, profits may double or triple but wages will remain as it was. Other factors of production will get highed return except labour. It had been in practice and it is still in practice. But it may be mentioned here that this problem is more acute in developing countries than developed countries as far as wages and hidden payroll is concerned. In developed countries like the U.S.A. the U.K. other European countries, Japan, Korea and rich Middle Eastern countries labourers are paid in such a manner that they can cope up with the expenditure. But labourers' in Bangladesh or in other developing countries are leading a subhuman life.

Trade Union Movement upto 1920

As long as industrial production remained in the hands of a number of independent master craftsmen and artisans, each working in his own or with the help of a limited number of apprentices, journeymen or hired labour, a personal contact was possible between master and the working people. This was the age of cottage industries and small traders. But the growth of factory system and later on industrial revolution completely changed the profile of the industrial arena. Gradually large-scale production and distribution came up that required engagement of vast labour and use of huge capital. This new production system required installation of large number of complicated machines that needed repair and replacements. So this was an era of transition from handicrafts to large-scale production and distribution, i.e., from labour-intensive to capital-intensive techniques of production. Everything changed rapidly but the conditions of labours remained static. Capital began to flourish, industries started growing, production increased but wages remained static and other conditions went down. Moreover, expansion of industries made personal contact between worker and management impossible. So workers could not take up their legitimate grievances to the management. The only weapon available to the workers was the formation of trade unions. Trade union thus started as continuing defuses pacts among workers against the inroads made into their rights and standard of living exigencies of capitalist self-interest (Subramanian, 1967: 5).

The trade union movement in the subcontinent took its birth in the 19th century. It is the development of factory industry which began after the industrial revolution in the U.K. that created conditions favourable for the organisation and growth of trade unions (Karnik, 1981: 1). So unionism has been one of the first fruits of the industrial revolution all over the world (Subramanian, 1967: 4). After industrial revolution in the U.K. labour class infused a new zeal to the increased number of labourers to get united to form an organization, i.e., the trade union organization. And it had a tremendous impact on the rest of the world especially on her colonies—India was one of her big colonies.

The history of the evolution of trade union movement in India may be traced back to the earliest times when guilds and panchayats in every village settled disputes between their members and their masters (Giri, 1962: 1). Working class of that time used to take up their grievances to the panchayats. Panchayats tried to redress their grievance to their level best, and in turn members followed the code of conduct of it. The scenario of every stage of the Indian trade union movement was same in Britain. The first form of permanent organization among wage earners was the local trade club of the eighteenth century (Flenders, 1962: 9). But those trade clubs or guilds even in the U.K. or any other western countries could not discharge any such function as modern trade unions do.

In India too the position of the workers was same. The guilds used to discharge some fraternal functions such as the grant of small financial assistance and enforced its members to a strict apprenticeship and high quality workmanship. The majority of Indian guilds were concerned with social affair and economic matters as well. Guilds used to regulate commercial and service activities as well as fix up appropriate wages so that its members would not accept, or be paid lower wages. But the characteristic feature of the guilds was that there was no element of collective bargaining over wages and conditions of service between employees and workers as, it happens in modern trade union (Subramanian, 1967: 2).

It goes without saying that the trade union is the outcome of modern factory system. Factories had been established very late in Indian sub-continent in comparison to the U.K. and other western countries. The first cotton mill was established in 1851 in Bombay and the first Jute mill in 1855 in Bengal. This was the beginning of modern factory system in India (Mathur & Mathur, 1962: 12). But Mr. Kamruddin Ahmed differs regarding the year of establishment of Jute mills in Bengal and furnished some detailed information about the mill. He stated that one Mr. Auckland started the Wellington Jute Mill in Serampore, the first of its kind in India in 1854 (Ahmed, 1978: 4). He further mentioned that by 1876, a fairly large number of Jute mills had been established in Bengal on the banks of river Hoogly (Ahmed, 1978, 4). K.N. Subramanian also expresses the same opinion regarding the establishment of the mill. According to him the first Jute mill was established at Serampore in 1854 (Subramanian, 1967: 7). So it maybe said that the mill was established in 1854. Whatever may be the year these two mills may be treated as the milestone in the way of modern factory system in India. After setting up of these two mills, mills and factories began to establish in different places of the country, especially by the British entrepreneurs, who have grabbed political power and economy too. By that time colonial rulers had never the intention to industrialise the country in the truest sense. They did what they were supposed to do for their own interest.

Labour class began to grow gradually with the establishment of mills and factories in India. But the economic condition of the working class was not improved. They were unorganized, not sufficiently literate and financially solvent. They had no working hours. There were no rules of wages, welfare programmes and legal support. They used to get subsistence wage and led a sub-human life. Not only in Bengal, in other parts of India also the vast majority of workers got a wage which was not enough to provide them with the meanest necessity of life (Dutt, 1956: 388). S.A. Dange has given a painful picture of Indian working people. According to him:

> From 1850 to 1880 the working class in these factories was exploited most inhumanly and without pity. Arrogant Britishers and their natives irrespective of their religion, nationally, language or country in bleeding men, women and children in their slaughter house of capital. There was no limit on hours of work. Men women or children were herded in the dues of capital to work from 12,16,18 and even 23 hours per day. There was no Sunday, no holiday, no starting and closing time. And when they died or maimed in the machine, there was no value of their file or limb (Dange, S.A.).

Thus from above delineation, it is clear that with the production of coal in 1775, first tea cultivation in 1834, and the setting up of Assam tea company in 1839, coffee plantation in 1840, first establishment of cotton mills in 1854 and jute mill in 1855 and opening of railway in 1853, industrial activity in Bengal as well as in India began to grow and a new class of industrial labour emerged gradually (Ramanujam, 1986: 5-6).

In this juncture the labour movement in India may be said to have began in 1875, when few philanthropists under the leadership of Mr. Sorabjee Shapoorjee Bengalee started an agitation in order to draw the attention of government to the wretched conditions of the workers, especially women and children, in factories and to the need for legislation, but the progress of the movement was disappointing (Panadikar, 1933: 15-16).

The History of trade union movement of undivided India may be divided into three periods: (i) From 1875-1920, (ii) From 1920 to 1930, (iii) From 1930 to 1947.

The first division of the period of trade union movement of India has some logical basis. The very year 1920 may be treated as the demarcating line in the history of the movement. Union movement did not begin till 1918 in proper sense. Few organisations started their journey in the same year. International Labour Office in 1938 reports:

> "It was not until 1918, however, that the trade union movement can be said to have begun in India. In that year, industrial organisation was started among the textile workers at Madras, and the clerks' union and postmen union were also founded; since then the number or organisations has been rapidly increasing in almost all industries throughout the country" (ILO Report, 1938: 123).

Another eventful incident that engulfed the whole world in a state of mental agony and dissatisfaction was the First World War. It was the first experience of mechanical war which thundered the world civilisation. The devastating war ended in 1917 but the war had far reaching effects of it in every sphere of life. League of Nations had been closed and the U.N.O. came into being with new hopes and aspiration. World started afresh in a new dimension. People became more conscious about peace, liberty and freedom.

After the end of the First World War, labour unrest and the formation of unions became almost a common affair all over India. Several factors contributed to this. The war-ravaged economy was left for the working class and people as well.

Rising prices affected not only the industrial workers in the towns but also the rural workers. Along with this economic distress came political upheaval: the "Jalianwalabag Massacre' and the intensified 'swaraj' or 'Home Rule Movement'. The setting up of the ILO at Geneva in 1919 acted as a symbol of recognition of the workers right to organize. Another factor was the success of the Russian revolution which gave an impetus to the labourers and the leaders (D'Costa, 1963: 86) And finally, the year 1920 saw the establishment of the All-India Trade Union Congress (AITUC) as the central organisation of all the unions of this organisation enabled the Trade Union Movement to emerge as a strong and stable force (Karnik, 1981: 16). This was also a great achievement to working class and it acted as a stimulus to the labour movement. Trade union organisations got a strong legal and secured platform as well. It can be treated as a milestone of the trade union history. So we can take the very year 1920 as demarcating line and dawn of an era.

All these events and incidents made the labourers more conscious than before. Gradually they began to think that they should not be neglected, as they are the inevitable part of the production and distribution system. Among the four factors of production they are not the least important. Capital is not the only factor of production. Labour is as important as capital. And if they are united they can challenge their lot and even they can change the system as a whole as it happened in Russia. Gradually they could consider labour as a strong class which capital and their puppet country administration can not withstand so easily. So class-consciousness began to develop gradually. It helped the labour to be united more strongly than even before.

With the establishment of some mills and factories in the 1870s some labour organisations began to grow. By 1879 there were 56 cotton mills employing 43,000 workers and 22 jute mills employing 27,494 workers. Due to the opening of railway in 1853 slow rise of collieries got speed and momentum which resulted in 56 mines in Raniganj area by 1879-80 (Subramanian, 1967: 7-8). In the next twenty years rapid industrial growth took place in India. So by the end of 19th century industrial activity was built around tea and coffee plantations, coal, cotton and jute industries. At the close of the 19th century there were 144 cotton mills and twines of over 77 million were exported. The number of jute mills was nearly 30, out of which 26 were in and around Calcutta. The coalmines employed about 43,000 workers who produced nearly 3 million tons of coal. Tea plantation industry had also tremendous expansion and prosperity at the end of the previous century. Iron and steel industry was established in Jamshedpur by Tata in 1907 (Ramanujam, 1986: 6). So number of people increased rapidly with the increased number of industries. It goes without saying that labour is the outcome of industries and trade unions is the outcome of labour. So without the growth of labour class the existence of trade union cannot be thought of.

Within two decades of the establishment of factories in India, the ludicrous condition of workers drew public attention and criticism. Major More, inspector in chief of Bombay cotton department, wrote in his Administration Report for 1872-73 that the hours of work of cotton mills had not been subjected to any regulation but as that women and children had been employed in large numbers the work was fatiguing, it was desirable to regulate at least the hours of labour for them. Major A.J. Ballard, Mint Master, Bombay also drew the attention to the necessity of factory Act for them. The Secretary of State for India took notice of the report and in 1857 drew the attention of the Bombay Govt. to the need for action (Subramanian, 1967: 8-9). Apart from this in 1875 few social reformers and philanthropists under the leadership of S. S. Bengalee started an agitation against the appalling condition of factory workers (Giri, 1962: 2). As a result, first factory commission was appointed in Bombay in 1875. The commission recommended at least a simple legislation for the protection of workers.

England sensed the precarious conditions of the factories and their interest at state raised their voice for the introduction of the Factories Act in India. On 30th July, 1875 Earl of Shaft Beery stated—we must bear in mind that India has the new material and the cheap labour; and if we allow the manufacturers there to work their orperatives 16 or 17 hours and put them under no restriction, we are giving them very unfair advantage over the manufacturers of own country and we might be undersold even in Manchester itself, by manufactured goods imported from the east (Ahmad, 1978 : 5).

Second factory commission appointed by this time met in 1884, where Indian sympathizers had arrived on the scene. A person who played a notable part in advocating the cause of labour and in supporting progressive factory legislation the early days was S.S. Bengalee, a member of both the 1884 and 1890 commission. Mr. Bengalee was not a trade union leader or representative of workers, but just an active sympathizer of them. M.N. Lokhande appeared as a witness before the 1884 commission, advocating the cause of labour (Subramanian, 1967: 30). The recommendation of the Indian Factory Commission (1884) was under consideration of the Government, chief inspector of factories in the U.K. alleging miserable conditions of work of Bombay, Manchester chamber of commerce moved for extending English Factories Act to India. The demand of two members in the British Parliament in favour of factory legislation in India and the resolution adopted at the Berlin International Labour Conference in 1890 compelled the Government for the appointment of the Indian factories commission in 1890 (Sharma, 1971: 50).

In 1877 the first strike over wage rates was recorded at the Empress mills at Nagpur, and few other strikes were also recorded in the eighties (Giri, 1962:2). S.S. Benglee, started first agitation for factory legislation. He drafted a bill and sent it to the Governor of Bombay in April, 1878. Obstinate resistance and organised opposition was put up by the Bengal and Bombay Chamber of Commerce, the Mill owners association,

Bombay and Calcutta traders association, the British India Association, the Bombay and Calcutta traders association, the British India Association, the Bombay mill owners Association, managers of different mills and a good number of private individuals for not introducing such legislation. The draft bill (1879) of the Government of India was originally based on Bengalee's Bill, but in the face of opposition from the business and manufacturing class it was very much whittled down (Sharma, 1971: 48). Despite so many obstacles and agitations the Government of India introduced a bill in the central legislation in 1879. The bill became an Act in 1881 (Ramanujam, 1986: 8). This was the first Factories Act. In India, though it did not come up to the expectations of labour. Whatever may be the reaction of the labour class as first factories Act it was a landmark in the manufacturing history, was also significant to the emergence of the working class' point of view in subsequent years, and ultimately to the birth of the labour movement in India.

Factories Act 1881 was made applicable to all manufacturing premises using mechanical power and employing 1000 or more hands and labour working for more than four months in a year. This Act also made provisions for limiting the working hours of child labour, prohibiting employment of women in night shifts and appointment of factory inspector for the first time of its kind (Ramanujam, 1986: 8). During the said period (1875-1920) trade union activity took place haphazardly which were not in western variety. The emergence of large-scale industrial and business organisation appeared very late in India in comparison to western world, union activity also started lately. Union activities of the said period were irregular, isolated and sectarian in nature. Those organizations were friendly and social welfare type especially in its early period of development.

In this sensitive situation the first labour leader in India Mr. Narayan Meghajee Lokhande, started his career as a factory worker and devoted his whole life to the labour movement. In

1884 he organised conference of the workers in Bombay factories for representing their grievances to the commission which had been appointed just before, and their memorandum was duly considered by the commission (Panadikar, 1938: 166). The first recorded case of collective representation of their claims by Indian workers was in 1884, when a conference of workers was called in Bombay for the purpose of drawing up a memorial to the Factory commission then sitting in that city (ILO Report, 1938: 123). But as no action was taken by the Government, the workers of the Bombay cotton mills sent a petition to the Governor General in 1889 reiterating and appealing for protection (Panadikar, 1938: 166).

M.N. Lokhande, organised a meeting of 10,000 workers in Bombay on 24th April, 1890 and presented a memorandum containing demands for reducing the hours of work, weekly rest day, mid-day recess and compensation for injuries. In response to these demands mill-owners of Bombay granted a weekly holiday. Encouraged by this success, Lokhande organised the Bombay Mill Hands Association in the same years, of which he was elected president (Giri, 1962: 2). This was the first labour association in India. In order to ventilate the aspirations and grievances of labourers a labour journal 'Dinabandhu' was launched. Though The Bombay Mill Hands' Association cannot be considered a trade union in real sense of the term. It did not exist as an organised body. The association had no role of membership, rules and regulation or funds (Mathur, 19:16). Whatever it might be it was the first organisation of workers and may be regarded as the foundation of trade union organisation in India.

Though the first Factories Act was passed in 1881, the Act was inadequate and disappointing. But the condition of workers did not improve. By this time another commission was appointed in 1890. Therefore another representation was submitted to the Government in 1890. This representation also reiterated the demand of 1884 and was signed by about 17,000 workers (Mathur & Mathur, 1962: 13). Mr. S.S. Bengalee was appointed as a general member and Mr. Lokhande as a local member for the Bombay Presidency on the Labour Commission of 1890 (Panadikar, 1938: 166). Later on new Factories Act was passed in 1890 which introduced several amendments. In this regard ILO report said –

> In 1890 a union, called the Bombay Mill Hands' Association, was organised, but this was loose combination rather than a corporate body, as it had neither a definite constitution nor paying membership (ILO Report, 1938 : 123).

Several strikes were recorded in the various parts of the country between 1882 and 1890. The workers of Great Britain also extended their moral support to the Indian workers. The textile workers of Lancashire organised a demonstration before the secretary of state for India demanding introduction of proper factory legislation in India (Ramanujam, 1986: 8). Two strikes took place in Bombay in 1894, but of little consequence. The first big strikes of mill operatives were recorded at Ahmedabad in the first week of February 1895. The Ahmedabad Mill-owners' Association decided to substitute a fortnightly wage system for a weekly one which had been in force since 1869. This forced the weavers, over 8,000 in number to leave work, but the strike was unsuccessful (Mukhtar, 1935: 13).

Indian labour being dutiful and hard working, worked 12 to 14 hours, sometimes 18 hours despite of the consistent indifference to their necessities of life. This condition made them frustrated and frustration brings violence. But Indian labour did not choose the violent way. They went for peaceful strike to ventilate their grievances and resentment.

On 11th May, 1897, the Mill owner's Association, Bombay, decided to discontinue the daily payment of wages. This led to several strikes which were resulted in resuming a partial continuance of daily wages (Mukhtar, 1935: 14). Labour organisations during those days were unorganised and few workers were interested into membership. In such a situation the first labour organisation in India was the Amalgamated Society of Railway Servants of India and Burma, which came into existence in 1897 and was registered under the companies Act. Early in the last century isolated unions were also started in several places, such as printers' union in Calcutta in 1905, the postal union in Bombay city in 1907, and the *Kamgar-Hitwardhak-Sabha* (Workmen's welfare Association) also in Bombay city in 1910 (ILO Report: 123).

Political leaders of this subcontinent have played a vital role uniting the labourers to form an association of their own. Political front had its reciprocal interest, as at the time of need workers association will respond to their call and simultaneously they could respond back when workers interest will be at stake. Politically British India was passing through a crucial time, as unrest prevailed all around. In this state of political situation the Swadeshi movement that added momentum in Bengal in 1905 grabbed the opportunity to be involved into the workers political activities. They were directed to create trouble for the big employers including Government (Subramanian, 1967 : 34).

Strikes took place in many other places all over India in early 20th century. A strike was recorded in the workmen in the press and machine section of the Madras Government Press in 1903 against overtime work without payment. The strike continued for about six months and after much suffering they returned to duty. In September 1905, the employees of the Government of India Press at Calcutta went on strike. Their main grievances among others, were-(i) non-payment for Sunday and Gazetted holidays, (ii) imposition of irregular fines, (iii) low rate of overtime pay, and (iv) the refusal of authorities to grant leave on medical certificate (Mukhtar, 1935: 14).

But the result of this strike was quite unsatisfactory. Seven ring leaders were dismissed and workers achieved very little out of this one month long strike. Another strike occurred at Samastipur Railway workshops in Bengal in 1907. The workers resumed work after six days, only when the authority granted a famine allowance (Sharma, 1971: 62).

Lokamanya Tilak was sentenced to six years' rigorous imprisonment in 1908 and so the workers of Bombay went on a six-day political mass strike (Subramaniam, 1967: 34). Here it becomes vivid why the politicians were interested in organising and patronising the labour union. The entire population was almost against the British rule and naturally labourers were not the exception. And moreover a god number of the owners of factories and business were the Britishers or their close allies. So the labourers had a hostile attitude towards the administration and considered the politicians of this country as friends. Naturally any oppression on the political leaders was considered as anti—liberation and anti-people activity.

The all-India telegraphic strike which took place in 1908 was successful. During this period (1907) some labour organisation like, Bombay Postal Union also started. It was in every sense a trade union. Indian Telegraph Association was formed in 1908 which used to publish one journal, 'Telegraph Recorder'. The Postal League started in 1902 (Sharma, 1971 : 66).

In 1910 they organised another strike demanding reduction in working hours. In the same year Bombay workers established two organisations, namely, Kamgar Hitawardhak Sangh and the Social service league. In 1911 the Government of India set up a commission for the reduction of working hours and it was reduced to 11 hours a day following the recommendations of the commission (Ramanujam, 1986: 10).

By this time the political situation was striving fast towards the achievement of total liberation of the country. Hot political situation gave impetus to the labour movement in the country. Moreover First World War broke out in 1914. The whole world had to face holocaust of war and India was not an exception especially her international trade sot a serious blow. In war time British policy towards Indians was softend to some extent. By and large it can be said that the trade union movement in this subcontinent during 1875 to the end of World War-I was essentially humanitarian. Some friendly societies were formed to look to the welfare of the workers. These unions were sporadic in nature (Sahoo, 1999 : 173).

The First World War ended in 1917. As an aftermath effects of war prices of essentials shot up like any thing, especially for the evil design of wholesellers and hoarders business became dull which caused the fear of retrenchment and unemployment for the labourers and the political front and labour front took new speed and momentum after arrival of Mr. Gandhi from South Africa. The Civil Disobedience Movement and Jallianwalabagh massacre turned the anger and furry of the people to a great extent against the foreign ruler. By this time workers revolution in Russia brought some new hopes to the toiling masses all over India. After Russian Revolution communism gained considerable ground in many countries of the world including India several strikes were recorded in the year 1917. By that time Mr. Gandhi started working for the labourers. Several meetings were organised by Mr. Gandhi in 1917. The stalwarts like Mr. Gandhi, Lokamanya Tilak, Annie Besant, and B.P. Wadia were very much active in the political as well as industrial area. Labour politics got a dimension to activate the participation of those leaders (Subramanian, 1967: 36).

The General awakening thus created among the working class led to a movement towards the organisation of labour. And as a result, the Madras labour union stood out very conspicuously in 1918 with Mr. B.P. Wadia as its first chairman (Mukhtar, 1935: 17-18). It was one of the fist well organised trade unions in the country developed according to modern concept. Mr. Wadia made a few concrete suggestions to Messrs Binney & Co. for the improvement of the conditions of their employees. But Messrs Binney & Co. did not reply. A largely signed petition from the workers suggested the alternative of a strike. In spite of giving reply Messrs Binney & Co. declared the first lockout in October 1918. Again on the morning of 27th November 1918, a second lockout was declared on the plea that manager and weaving master were assaulted by unknown hands. Finally the mill reopened on the 17th December 1918 with partial fulfilment of the labourers demand (Mukhtar, 1935: 19). Another strike by weavers was recorded in Ahmedabad in 1918 on bonus issue against the Mill Owners' Association. At a later stage Mr. Gandhi declared vow to fast till the matter was settled. The Mill Owners' Association had to accept the demand according to Mr. Gandhi's proposal (Mathur & Mathur, 1962: 17). An I.L.O Report says:

> In 1918 an industrial organisation was started among the textile workers at Madras, and clerk's union and postman's union in Bombay and the seamen's union in Calcutta were founded; since then the number of organisations has been rapidly increasing in almost all industries throughout the country. In this development of trade unionism, the existence of the international labour organisation has played an important part (ILO Report : 1938 : 123).

World war first not only brought huge profits to industrialists but also led to a sharp rise in prices. The consequent distress to workers, wages were not correspondingly increased, generated a series of strike waves in 1918-19 (Giri, 1962: 10). Madras became the champion in case of labour activity. Strikes were taking place one after another, not only in Madras but also dissatisfaction and unrest prevailed throughout the country. Anti-British movement was also mounting day by day. Sixteen strikes took place in the Madras Presidency and the demands of the workers regarding increase in wages, grant of bonus, rice allowance, reduction of working hours and extra holidays continued. Eight strikes were recorded in Bengal, Bihar, Orissa and Assam. Another general strike was recorded in Bombay where they gained an increase in wages (Mukhtar, 1935: 20).

Ten other unions were organised in 1919 of which five in Bombay, two in Madras, one each in Bengal, U.P. and Punjab. The Employees Association, Calcutta, the Seaman's Union, Bombay and M.S.M. Railway Union, Madras were the important unions that were organised during 1919 (Mathur & Mathur, 1962: 17). During the winter of 1919-20 again there was a major strike in the cotton mills of Bombay, involving 1.5 lakhs of workers. The factors responsible for this strike were longer hours of work, lack of amenities and facilities and victimization of workers by the supervisory staff (Giri, 1962: 10).

Trade Union Movement from 1920-30

The period from 1920 to 30 is marked as beginning and growth of organised and continuous trade unions, the seed of which had already been sown through the establishment of Madras labour union in 1918. By this time things had changed. Workers became more courageous with the help of the politicians. The Indian National Congress began to take interest in Trade Union Movement since 1919. The Trade Union Movement thus became an integral part of freedom movement (Sharma, 1990 : 90). By this time the all India Trade Union Congress (AITUC) was formed in the year 1920. Its first session was presided over by Lala Lajpat Rai who was then also president of the Congress (Ramaswami & Ramaswami, 1981: 89).

During the annual session of the Indian National Congress in 1921, the National Congress welcomed the formation of the AITUC in 1920 and in the Gaya Session in 1922, the Indian National Congress had also appointed a committee to assist the executive of the AITUC for the organisation of Indian labour (Sharma, 1990 : 93). The formation of AITUC was the direct result of the formation of the International Labour Organisation (ILO) in 1919. It was very much necessary to have a national central organisation of labour. Moreover, in order to nominate a delegate to the ILO, it was necessary to have a centre like that (Ramaswami & Ramaswami, 1981 : 89). Large number of trade unions was formed by 1920. Just after the establishment of AITUC 64 trade unions with a total membership of 140,854 were affiliated with it (Mathur & Mathur, 1962 : 17).

They had associated themselves more with the national movement. In this way Trade Unions inclined gradually with politics. Prior to First World War trade unionism was confined to a small section of workers; but after the war it went up to the whole working class. So the period from 1920 to 1930 was the period of growth of organised labour in India. Due to First World War there was acute price hike, profit maximisation to industrialists, growing political agitation with labour unrest that gave an impetus to trade union movement. But wages were not increased.

The period between 1920 and 1930 was very eventful. During this period many agitations and strikes and organisational development of trade union took place in India and Bengal as well.

The following table [2.13 (a) and 2.13 (b)] summarises the position of trade unions affiliated to AITUC in 1920.

It is marked onward the tendency of the spread of trade unionism with large number of strikes. Industrial relations were deteriorating fast. Mistrust developed amongst the workers due to the long deprivation and suppression and oppression. Though the attitude of management to some extent had changed by that time, the workers could not trust or depend on their masters so early.

International Labour Organisation estimated the total membership of Indian trade unions at 5 lakhs. But Dr. Punekar did differ regarding the number of workers. He opined that there were 125 unions with a membership of 250,000 (Mathur & Mathur, 1962: 18). Unrest was on throughout India in this period. The economic condition had worsened even further and nearly 200 strikes were declared all over the country in 1920. Important strikes during 1920 were the general mill strike in Bombay, the general spinners strike in Ahmedabad and a general strike in Borach and Sholpur. In Punjab the Railway workers went on strike. Sixty-two strikes were declared in Madras and particularly all industries were affected in Bengal, Orissa and Assam. Government of India started collecting information about strikes in 1921. In all 400 disputes involving 523,151 workers were recorded. The strikers' main demand was higher wages in 174 cases; bonus in 75 cases. The question was

Table 2.13 (a) : Unions Affiliated and Sympathetic to AITUC in 1920

	Number of affiliated and sympathetic unions	*Number of affiliated unions*	*Membership of Affiliated unions*
I. According to province			
Bombay	56	44	48,88
Bengal	5	1	2,505
U.P.	08	3	15,800
C.P.	6	2	128
Sind	2	1	128
Madras	16	8	3,559
Bihar	1	-	-
Punjab	9	4	70,253
Delhi	2	-	-
Indian States	1	1	1,600
Ceylon	1	-	-
Total	107	64	140,854

Table 2.13 (b) :

	Number of affiliated and sympathetic unions	*Number of affiliated unions*	*Membership of Affiliated unions*
II. According to Industries			
Railways	21	11	91,427
Textiles	12	9	7,719
Shipping	4	3	19,800
Transport	4	2	2,470
Chemical	7	6	856
Engineering	8	7	7,590
Posts and telegraph	15	5	1,685
Printing and papers	7	3	1,844
General	29	18	7,463
Total	107	64	140,854

Source : Mathur & Mathur, 1962: 18.

involved in 63 cases and other personnel's for leave etc. 162 strikes were declared in Bombay and 135 took place in Bengal (Mathur & Mathur, 1962: 19). So we see that there were waves of strike in early twenties of the last century. This proved that the workers became more powerful than before. Organisationally they were stronger as they had good number of labour unions and in 1920 AITUC came to action. Moreover they were getting low wage and very poor employee benefits and services programme. This was the major reason of dissatisfaction in the labour front.

In 1922, 278 strikes involving 435,434 workers were successful. In 127 cases increased wage was demanded, another 32 strikes the issue involved increased wages and in another 32 strikes the issue was hours of work. Bombay, Bengal, Bihar, and Orissa were the chief provinces of industrial unrest (Mathur & Mathur, 1962: 19). Next, in Ahmedabad serious strike was recorded on the 1st April, 1923. It affected 56 cotton mills, involved 43,113 working people which resulted in total time loss of 2,370,933 working days. 132 strikes were recorded in 1924, large numbers of which were unimportant, and short-lived. There were 56 disputes in Bombay alone, resulted in a loss of 7,559,401 working days; 25 percent of these strikes were settled in favour of the employees, 8 percent were compromised and the rest were failed. 134 strikes took place in 1925 and 33 percent of these were successful. In the same years serious type of strike broke out in Bombay against reduction in their wages. All the mills especially cotton mills were closed down completely on 2nd October 1925, and the strike was brought to an end after involving a loss of about 11 million working days (Mukhtar, 1935: 40).

Ideological difference and international outlook may exist among people. This difference of ideologies or opinion some times creates conflict. It was infused among the labourers especially after the success of Soviet Russia. Actually after the rise of Soviet system labour as a class of the world has been divided two blocks, and which was almost clear in the trade union congress in the Geneva international conference. India was not an exception to that; rather this division became apparent within a short time. Regarding the Indian scenario V.V. Giri commented:

> "By 1927 two distinct parties grew in the Indian trade union movement. They were called by some 'rightists' and 'leftists', and by others, Geneva-Amsterdam groups and Moscowvites" (Giri, 1962: 13).

Influenced by the international labour politics division of the working class of India became imminent. Division and disunity brings partial fulfilment of the objective. Moreover economic and social condition of the working class of India was wrecked. Disunity strengthened the hands of the capitalists and their close allied Government administration. But it was the reality that the conflict between the communists and the moderates surfaced in the late twenties. The communist believed in class-war and utilisation of the labour movement for the overthrow of the moderates. Moderates formed the Indian Trade Union Federation (ITUF) in 1929 (Mathur & Mathur, 1962 : 27). The first split took place at the tenth session of the AITUC held in Nagpur in 1929 (Sharma, 1997: 92). Meanwhile, the hardcore communities had left the AITUC to form the Red Trade Union Organisation named All-India Trade Union Federation in 1932 (Sharma, 19: 99).

By this time, in the middle of 1929 Royal Commission on Labour was appointed and the Commission submitted its report in 1931. The influence of the communists began to decrease after 1930. The failure of the general strike sponsored by them during 1929-30 produced harmful effects. Other strikes also proved ineffective in getting concessions for the workers. These failures made the workers distrust the communist leadership. There was also a vacuum of leadership as the old experienced communist leaders were involved in the Meerut trials. Moreover, the communists had lost credit as a result of the split in the AITUC and the very way to control the Congress. Finally the Civil Disobedience Movement engaged the public attention.

In this way Communists lost the public sympathy because of their aloofness from the political movement the country (Mathur, 19 : 30).

Thus, after 1930 the trade union movement in India got two splits which may be considered a disaster to the rising united movement against exploitation and deprivation. The trade union movement in India was going ahead in one platform, but from 1930s it went on in the divided way.

Trade Union Movement from 1930-1947

The growth of trade unions, however was not a smooth one, it had its ups and downs. Mohmmad Anisur Rahman said the period 1926-39 is characterized by three distinct features; viz., (i) the economic recession during the thirties, (ii) the formation of popular ministries in the Indian province, and (iii) conflict in the trade union leadership (Rahman, 1968: 45).

The country had to face an economic depression in the early 1930s. During 1929-34 the economic problem became very acute. Real war started after the war, i.e. socio-economic war, which happened in India. Side by side political problems also mounted up. The dominant trade union leaders were arrested in 1929 in connection with Meerut conspiracy case. Finally the situation was further aggravated by the serious split up of AITUC. Employers took the full advantage of the situation. They adopted a policy of retrenchment of workers on a large scale, imposed wage and started demoting employees to lower grades (Giri, 1962: 16).

Workers used the instrument of strike to get hold of their employers, but could not succeed in many cases, as the economic recession was on and the workers were divided. In 1932 we get signal of unity. Red Trade Union Congress, which was born due to a split in the Trade Union Congress was again united to the parent organisation (AITUC) (Giri, 1962: 17).

In 1930s the climate was unfavorable for the growth of trade union movement. Gandhiji's civil disobedience movement launched in 1930, had drawn the attention of the political leaders and the Government of India. The arrest of the prominent communist leaders in the Meerut conspiracy case and the failure of Bombay textile strike of 1929 also brought stagnation in the trade union activity (Giri, 1962 : 16).

In 1930s, 148 strikes were recorded but only 196,301 men were involved in them. The loss of working days amounted to 2,261,731 and the workers failed to gain any concession in 61 percent of the strikes. In the same year in protest of Mr. Gandhi's arrest ship loading work was delayed in the port of Rangoon. And simultaneously the coolies, generally Telegus, struck work on the demand of higher wages. They were successful to enhance the wage from 1-12-0 instead of 1-8-0. But when they joined, serious rioting followed with the Burmese labourers who had acted as strike breakers. Later on, a conciliation board effected a settlement based on the understanding that the work on the ships would be equally divided between Burmese and Telegu workers (Mukthar, 1935: 73).

In the year 1931, 160 strikes occurred in which 203, 008 workers took part and 2,408, 123 man days were lost. 118 strikes took place involving 128,099 workers and 1,922,437 man-days lost in 1932. Only in 36 percent of the strikes the workers got some positive results. The number of strikes in 1933 was 146 wherein 264,938 workers participated and 2,168,961 man-days lost, gained in only about 31% of the disputes (Mathur & Mathur, 1962 : 25-29).

Prolonged strikes took place in Bombay, Sholapur and Nagpur. The strike in Bombay which lasted from April to June 1934 involved 90,000 workers. The strike in Sholapur lasted from February to May and from May to July in Nagpur. A total of 4,775.59 man-days lost were in that year. Due to unemployment and low wages, strikes continued between 1937-39. Among these were the general textile strikes of Kanpur (1937-39), the jute workers' strike at Calcutta (1938), the Digboi oilfields strike (1939) and the U.P. glass bangle factories strike (1939) (Subramanian, 1967: 63).

Business began to recover from depression since 1934 and with it trade union activity also revived. The ruthless attitude of the employers towards unions began to change gradually and the hostility was also decreasing. The employers needed to realise that things were changing globally with the change of time. So a healthy situation was emerging that followed the unity of federations on the national level.

Table 2.14 below shows that the year for which membership was in peak was in 1934-35. Though there were, in 1935-1936, 23 more registered unions, the membership figure declined. The number of women members of registered unions which made returns rose from 1,166 in 1927-1928 to 7,309 in 1935-1936 (ILO Report, 1938 : 124-125).

Table 2.14 : Number and Membership of Registered Unions in India

Year	*Registered unions*	*Registered unions submitting return*	*Total membership*	*Average membership per registered union*
1927-1928	29	28	100,619	3,594
1928-1929	75	65	181,077	2,86
1929-1930	104	90	242,355	2,693
1930-1931	119	106	219,115	2,067
1931-1932	131	121	235,693	1,948
1932-1933	170	147	237,369	1,615
1933-1934	191	160	208,071	1,300
1934-1935	213	183	281,918	1,557
1935-1936	236	205	268,326	1,309

Source : ILO Report, 1938.

We see a good sign of unity in Indian trade union movement in 1935, which is essential for the development of trade union movement. The communists dissolved the Red Trade Union Congress and rejoined the AITUC in 1935. Thus, the AITUC once again became the sole representative of the organised labour (Sharma, 1990: 99).

A new constitution was adopted in 1935. Under the constitution, labour representatives could get themselves elected through labour or trade union constituencies (Mathur & Mathur, 1962: 29). This decision caused harm to some extent because the dissension amongst the leaders was affecting the interest of the workers, therefore the efforts of unity was hampered. Another important event took place in 1937 in the political history of the sub-continent. Under the provisions of the Government of India Act, 1935, general election was held in 1937. In the election leaders and people of this land contested the election, and popular ministries were formed in different provinces of India (Rahman, 1968: 46). This change in Government machinery in the states created hopes and aspirations among the workers. They thought that their conditions would be improved. The popular ministries took some time to implement their programmes, but workers could not wait for long, i.e., there was a gap between expectation and reality.

The delay in realising the workers' expectations resulted in a recurrence of labour unrest all over the country. There were strikes in almost all the major industrial centres, for example Bombay, Calcutta, Ahmedabad, Kanpur and Madras, for higher wages and restoration of wage cuts on the ground that there was prosperity in trade and business. Again they were dissatisfied regarding their working and living conditions (Giri, 1962 : 19).

The general and big strikes occurred between 1937 and 1939. Among those were the general textile strikes at Kanpur (1937-39). The jute workers strike in Calcutta (1938), Digboi oil fields strike (1939) and the UP glass bangle factories strike (1939) were recorded accordingly. The number of registered trade unions increased from 241 in 1935-36 to 555 in 1938-39. The number of registered trade unions between 1939-40 and 1944-45 increased from 666 to 865. Table 2.15 below shows the progress of trade unionism between 1934 and 45.

Table 2.15 : Registered Trade Unions During 1934-1945

Year	*Number of Registered trade unions*	*Number of unions submitting returns*	*Membership of unions submitting returns*	*Number of women Members*
1934	191	160	208.071	-
1935	213	183	284.918	-
1936	241	205	268.246	-
1937	271	228	261.047	-
1938	420	343	370.112	-
1939	562	394	399.159	-
1940	667	450	511.138	18,612
1941	727	483	513.832	19,417
1942	747	455	573.520	17,094
1943	693	489	685.299	25,972
1944	761	563	780.967	26,866
1945	865	573	889.338	36,315

Source : Subarmanian, 1967: 63-68.

The period from 1939 to 1947 was crucial for the Indians, Government and the rest of the world. The holocaust of the second Great War in 1939 and its prolongation till 1944 and the intensity of liberation movement in India had a tremendous impact on the attitudes of the Indian labour. During this period they nearly turned against India and the British Government.

The economic consequence of war is of more concern to us. The war brought about considerable changes in the economic spheres of life. There was great shortage of essentials, created by the hoarders and wholesale traders. Production and distribution were also hampered due to transport bottlenecks. As foreign goods were not available in the market due to shortage of shipping during the war, provided an indirect protection to industries to a considerable extent. As a result Indian industries began to grow.

This was the period when, willingly or unwillingly, the Government made efforts to maintain industrial peace by bringing the employees and the employers to the conference table. Joint consultations with the representatives of the employees and the employers were undertaken; and in 1942, the first tripartite labour conference was organised by the Government (Giri, 1962: 23).

During this period of economic activities money wages increased, but at the same time cost of living went up to a great extent. So the increase of money wages could not cope up with the rising prices, and as a result, economic condition of workers became worse than ever before. So dissatisfaction and unrest among the labour force and common people continued. The Earnings of the workers and profits earned by the industrialists during the year 1939 is sufficient to understand the degree of exploitation. The real earrings of the workers during 1940-45 are shown below in table 2.16.

Table 2.16 : Index of Real Earnings of Workers During 1940-45 (Base 1939)

Year	*Index of Earnings*	*All India Consumer price index*	*Index of Real Earnings*
1940	105,3	97.0	108.6
1941	111.0	107.0	103.7
1942	129.1	145.0	89.0
1943	179.6	268.0	67.0
1944	202.1	269.0	75.1
1945	201.5	269.0	74.9

Source : Mathur and Mathur, 1962: 38.

The following table will show the index of real profit during 1940-45.

Table 2.17 : Index of Real Profits (All Industries) During 1940-45 (Base 1939)

Year	*Index of Profits*	*All India Consumer price index*	*Index of Real Profits*
1940	138.0	97.0	142.3
1941	187.0	107.0	174.8
1942	221.8	145.0	153.0
1943	245.0	268.0	91.4
1944	238.9	269.0	88.8
1945	233.6	269.0	86.8

Source : Mathur and Mathur, 1962: 39.

The post war trade union development will be clear from table 2.18 below.

Table 2.18 : Growth of Trade Unionism During 1944-45 to 1947-48

Year	*No. of Registered trade unions*	*Registered Number of unions submitting returns*	*Membership*			*Average*
			Men	*Women*	*Total*	
1944-45	865	573	8,53,07	36,31	8,39,38	1,552
1945-46	1,007	585	8,25,46	38,57	8,64,03	1,480
1946-47	1,833	998	12,67,16	64,79	13,31,96	1,335
1947-48	2,766	1,620	15,60,63	102,29	16,62,92	1,026

Source : Mathur and Mathur, 1962 : 44.

So far the previous pages have discussed the development of trade union movement up to mid-1947. The British Raj in India came to an end with the division of the country on August 14, 1947 and the two independent states emerged on the world map—Pakistan and Indian union. So the movement rolled over in new dimension with divided scenario.

Summary

In this chapter an attempt has been made to discuss the state of cottage industry and small trade of India in the medieval period and her golden heritage of indigenous products in home market and abroad. Later on the country gradually switched over from cottage industry to the factory system in the mid 18th century, and that resulted in the birth of industrial labour. From the very inception of the factory system labourers were being deprived of minimum wages and employee benefits and service programmes. Deprivation and suppression induced the working class to be united to form trade union organisation. Mr. M.N. Lokhande formed the Bombay Mill Hands Association in 1890 and it was the first attempt to form trade union. This attempt would be treated as the milestone in the history of the labour movement.

Labour leaders and the politicians had to undergo a long struggle to give a solid stand to the trade unions. First federation of trade unions was formed in 1920 and it was the first of its kind. Due to continuous pressure and joint movement by the workers and political persons, different labour commissions were formed for the improvement of the socio-economic conditions of the working class. And finally Trade union Act of 1926 was passed. This Act gave the fundamental right of freedom of association which was denied by the employer. Thus trade union organisations got legal recognition.

After the Second World War political and ideological conflict of left and right intensified globally which had its influence in India too. Division and disunity became apparent in labour front, caused split up within unions and federations. Divided strength strengthens the hand of the employer. In all, we have discussed the emergence of labour with the industrial development of India, socio-economic condition of labourers and trade union movement from the very inception to 1947 in different phases.

NOTES & REFERENCES

1. In textile factories, excessive hours are frequently found in cotton mills. In all jute mills also weavers are employed for excessive hours. In Ahmedabad the average of the whole year was approximately 12 hours and 7 minutes, actual working time in the longer days the operatives come to the mills about 5-30 a.m., and leave at 7-30 p.m., giving 13.5 hours working time, the shortest working day is approximately 11 hours. In Bombay the scenario was the same. In some cases, mills worked from 5 a.m. to 19-35 p.m. of over 15 hours actual work. Later on the Bombay Mill Owner Association, at a meeting held in August 1905, passed a resolution in favour of an average 12 hours day. The mills fitted with electric light generally began work about 5-30 a.m. and continued with usual half an hour interval about noon, to 7 p.m. A few mills begin work about 5-40 a.m. working till 7-15 or even 7-30 p.m.; in one or two others hours are 6 a.m. to 8 p.m. - giving 13 hours actual works all the year round. The mill's working hours begin about a quarter of an hour before sunrise, and stop about a quarter of an

how after sunset; the adverse working time, over the whole year, is approximately 12 hours and 8 minutes; the shortest working day is about 11 hours and 10 minutes, from 6-55 a.m. to 6-35 p.m.; and the longest from 5-45 a.m. to 7-33 p.m.-giving 13 hours actual working time.

One mill worked a 10 hour day only, from 8 a.m. to 6 p.m.; another worked the spinning department for 12 hours continuously from 7-30 a.m. to 7-30 p.m., the weavers working from 6 a.m. to 7-30 p.m.; a third mill daylight hour day, without the midday interval; a forth worked from 8 a.m. to 8-30 p.m., also without an interval; and a fifth from 5 a.m. to 8 p.m., with an interval of 2 hours in the middle of the day in Calcutta. The jute mills worked nominally, from 5 a.m. to 8 p.m. without stop. The workers, except in the weaving and sewing are divided into a number of shifts, the hours of work of each shift being arranged so that each worker had an interval of at least two hours – or else is entirely free – at the time when he took his principal mill. Where no longer interval is given–i.e. where the worker ate his food before beginning of work, or after finishing it –each operative had a rest interval of half an hour or so during the term of his shift. The shifts are so arranged that no operative worked more than 11 hours a day.

The usual hours of work were from 6 a.m. to 6 p.m. in the cold season; and from 5-30 a.m. to 7 p.m. in the hot weather. The two largest mills worked on an average, about 11 hours a day only throughout the year; largest day did not exceed 11 hours in Madras town. All the mills in Madras presidency worked in the daylight only, the longest day being from 5 a.m. to 6-45 p.m., and the shortest from 6 a.m. to 6 p.m. In no case do the longest hours' exceed 12 hours, and generally the hot weather day does not exceed 12 hours.

All the mills worked daylight hours only in central province. The average working time for the whole year was approximately 12 hours 3 minutes and the longest day in any mill was from 5-25 a.m. to 7-20 p.m. –giving 13 hours 25 minutes actual work. The shortest working day was approximately 11 hours. In the cold season the mills actually worked 13 hours, or even 14 hours a day, beginning as early as 6 a.m. and stopping as late as 9-30 p.m. The hot weather hours' are usually from 5 a.m. to about 7-30 p.m. Some mills allowed interval of one hour, others an interval of half an hour or three quarters of an hour.

The mid-day interval or one hour was peculiar in Punjab. In Amritsar and Lahore the actual working hours varied from 11 to 13 hours in the cold weather, and average 13 in the hot season. In Lahore one mill worked similar hours, but the other occasionally worked up to 9 p.m. and had an average 13 hours actual working day throughout the cold season (*Report of the Indian Factory Labour Commission:* 1908).

2. *'Malangis'*—Salt was produced by a class of people known as *malungis* whose profession was by and large hereditary (Awal, Iftekar-ul, 1997).

Ahmad, Kamruddin, (1978), *Labour Movement in Bangladesh,* Inside Library, Dhaka.

Ahmad, M.S., (1969), *Year Book of Jute,* 1967-68, Part-I, Asian Trade Publications, Dhaka.

Anstey, Vera, (1952), *The Economic Development of India,* Longmans Green and Co., London.

Awal, Iftekar-ul (1997), State of Indigenous Industries, *History of Bangladesh,* Vol. 2, Sirajul Islam (ed.), *op. cit.*

Bengal District Gazetteers, Calcutta, 1907.

Bengal Distric Gazetteers, Jalpaiguri, Allahabad, 1911.

Bhowmik, Sarit, (1981), *Class Formation in the Plantation System,* Peoples Publishing House (P) Ltd., New Delhi.

Bose, Nemai Sadhan, (1969), *The Indian Awakening in Bengal,* Firma K.L. Mukhopadhyay, Calcutta.

Boserup, Easter, (1970), *Women's Role in Economic Development,* George Allen and Unwin Ltd., London.

Broughton, G.M., (1924), *Labour in Indian Industries,* Humphrey University Press, London.

Census of India, (1901), Vol. 1, Part-1, Report.

Census of India, 1911.

Census of India, 1921, Vol. 1, Part-1, Report.

Chakrabarty, Dipesh (1996), *Rethinking Working Class in History : Bengal 1890-1940,* Oxford University Press, Bombay.

Chowdhuri, Sushil, (1975), *Trade and Commercial Organisation in Bengal (1650-1720),* Firma K.L. Mukhopadhyay, Calcutta.

Chowdhuri, Sushil (1997), *European Companies and Export Trade, History of Bangladesh,* Vol. 2, ed. Sirajul Islam, Asiatic Society, Bangladesh, Dhaka.

Chowdhury, S.C., (1987), *Social, Cultural and Economic History of India,* Surjeet Publications, Delhi.

D'Costa, Ronald, (1963), *The Role of the Trade Union in Developing Countries (A Study on India, Pakistan and Ceylon),* Havaux, Nivelles.

Dange, S.A., (1925), *On the Indian Trade Union Movement,* A Communist Party of India Publication, Bombay.

Dutt, Rajni Palm, (1970), *India Today,* People Publishing House, New Delhi.

Dutt Romesh, (1902), *Economic History of British India,* London.

Dutt, Romesh, (1956), *The Economic History of India (Under Early British Rule),* Routledge and Kegan Paul Ltd., London.

Flenders, Allen, (1962), *The Trade Unions,* Hutchinson & Co. Publishers Ltd., London.

Giri, V.B., (1962), *Labour Problem in Indian Industry,* Asia Publishing House, Bombay.

Gosh, Tushar Kanti, (1978), *Tea Gardens of West Bengal,* B.R. Publishing Corporation (Division of D.K. Publishers Distribution (P) Ltd., Delhi).

ILO Report, (1938), *Industrial Labour in India,* Geneva.

Karnik, V.B., (1981), *Trade Union Movement and Industrial Relations,* Somaiya Publications Pvt. Ltd., Bombay.

Kling, Blair, B., (1966), *The Blue Muting, The Indigo Disturbance in Bengal 1859-1862,* Philadelphia University of Pennsylvania Press, USA.

Mathur, A.S. and Mathur, J.S., (1962), *Trade Union Movement in India,* Chaitanya Publishing House, Allahabad.

Mukhtar, Ahmad, (1935), *Trade Unionism and Labour Disputes in India.* Longmans Green & Co. Ltd., Bombay.

Panadikar, S.G., (1933), *Industrial Labour in India,* Longmans Green & Co. Ltd., Bombay.

Rahman, Anisur, (1968), *Trade Unions in Pakistan : An Evaluation of Their Problems and Development,* unpublished dissertation of Degree of Master of Science (Social Science) of the University of Edinburgh.

Ramanujam, G., (1986), *Indian Labour Movement,* Sterling Publishers Ltd., New Delhi.

Ramaswami, E.A. and Ramaswami, Uma, (1981), *Industry and Labour : An Introduction,* Oxford University Press, Delhi.

Report of the Indian Factory Labour Commission (1908), Vol. 1 - Report and Appendices, Simla.

Report of Administration of Bengal, 1929-30 (1931), Government of Bengal, Bengal Secretariat Depot., Calcutta.

Report of Administration of Bengal, 1932-33 (1934), Government of Bengal, Bengal Secretariate Depot, Calcutta.

Review of the Trade in India in 1942-43, Office of the Economic Adviser, Government of India, Published by the Manohar Publications, Delhi.

Robert, J.A. (1904), *Review of the Trade of India in 1903-1904,* Office of the Superintendent of Government of India Calcutta.

Saha, Panchanan, (1978), *History of Working Class Movement in Bengal,* Peoples Publishing House, New Delhi.

Sahoo, Basudeb (1999): *Problems and Prospects of Labour Movement in India,* Rawat Publications, Jaipur.

Sarkar, Jagadish Narayan, (1987), *Mughal Economy : Organisation and Working,* Naya Prakash, Calcutta.

Subramanian, K.N., (1967), *Labour-Management Relations in India,* Asia Publishing House, Bombay.

Sen, Sunil Kumar, (1981), *An Economic History of Modern India,* Progressive Publishers, Calcutta.

Sharma, G.K., (1971), *Labour Movement in India : Its Past and Present,* Sterling Publishers (P) Ltd., New Delhi.

Sinha, J.C., (1927), *Economic Annals of Bengal,* Macmillan and Co. Ltd., London.

Sinha, N.K., (1961), *The History of Bengal,* Vol. 1, Firma K.L. Mukhopadhyay, Calcutta.

Sinha, N.K. ed. (1967), *The History of Bengal (1757-1905),* University of Calcutta.

Sirajuddin, A.M., The Condition of the Salt Manufacturers of Bengal under the Rule of East India Company, *Journal of the Asiatic Society of Bangladesh,* Vol. XVIII, No. 1, 1973.

Statistics of British India, 1921, Vol. 1, Commercial Statistics.

Statistics of British India, 1922, Vol. 1.

The Indian Cotton Textile Industry (1940 Annual), September 1940.

The Indian Factory Commission of 1908.

The Statesman Year Book, 1918, ed. Sir John Scott Ketti Macmillan and Co. Ltd., London.

Vakil, C.N., Bose, S.C. and Deolalkar, P.V., (1931), *Growth of Trade and Industry in Modern India,* Longmans Green and Co. Ltd., Bombay.

Chapter - 3

GROWTH OF THE MOVEMENT DURING 1947–1971

Introduction

Trade Union movement started in a new dimension in 1947 as the country was divided into two independent nations –India and Pakistan. They started their activity into two separate environments and under separate Governments, though they had the common experience of fighting against capitalism and foreign master. East Bengal (the region of East Pakistan) was industrially backward in the British regime, so the trade union movement in the areas forming Pakistan was less strong than territories falling within Indian Union. Moreover, what it achieved regarding union activity that credit went to 'All India Trade Union Congress' (AITUC). The AITUC was instrumental in giving the trade union movement in East Bengal its early backbone (Ahmed, 1978 : 31).

There are some salient features of trade union movement of Indian subcontinent which has definite effect on trade union movement of Pakistan and Bangladesh as the later two geographical territories have originated from undivided Indian union. There were some common features among them.

During British period workers were very much insecure as they were ill paid and had no support either from Government or their masters. Moreover, labourers were treated as low class people in the society though they were the integral part of production process. Almost the entire labour force was unskilled. And unskilled labour always receives low wage. Skilled labours were seldom found amongst the natives. There were strong reasons behind it.

The indigenous industries of Bengal as well as India had been destroyed in a planned way in the aim of introducing British manufactured goods in the early 19th century. As a result, large number of artisans became unemployed and they had no option but to join agricultural activities (Mukherjee, 1951: 1). But when factory system had been introduced in the middle of 19th century landless work force and displaced villagers, as well as urban craftsmen were recruited as industrial labour (Weatheford, 1962: 18-19). And management could not arrange training for those newly recruited workers. Workers used to follow their predecessors. Actually the foreign or native owners or the management did not try to give birth to an industrial labour class in India and Bengal. But in western countries including the UK, Industrial Revolution took place during 18th century, the change over from cottage to factory production was more or less natural one with no gap in between (Rahman, 1969: 157). So craftsmen switched over directly from the cottage to the factory production.

Thus workers were most unskilled as they were coming from agricultural background. There was acute shortage of skilled workers. Moreover management's attitude too was not favourable. Another reaons was that workers used to switch over from one mill to another for better wages and better working conditions. It is an economic truth that productivity is always co-related with living and better working conditions. So employer and employee should put their heads together to create better working conditions and increase productivity (Malik, 1954 : 13). Recruitment was also defective. In all the factories workers were recruited through middlemen (i.e. jobbers), in local term labour *Sarder*. These middlemen used to take bribe from the workers. So the workers were more loyal to the middlemen than to the employer. These middlemen very often used to change the workers from one factory to another. These middlemen were also responsible for high turnover which led to the unstable work force and industries as well (Siddique, 1989 : 86-87). Moreover frequent change of jobs was one of the impediments to be skilled and experienced workers. Unskilled workers had little idea about trade union or no experience to be united to form any association. These workers had a very little knowledge about collective bargaining process. That is why they had to depend on a group of leaders known as 'outsiders'. But when we look at the UK or other western countries we can see a different picture.

The skilled workers first took up trade union movement in the western countries. The skilled workers were found to be associated with 'trade clubs' or societies (first with the trade guilds or the association masters, journey men and apprentices or later with separate societies organised independently by the workers) even before the industrial revolution (Cole, 1962: 11-13). In many other European countries industrial revolution took place in 18th century. Those countries switched over from cottage industries to factory production and there was no gap. There was natural and continuous change over from small to bigger factories. So they had checkered experience. Due to this continuous process supply of industrial labour and trade union leaders emerged through a gradual process. But the situation in the Indian subcontinent was completely different. Workers had no legal support, there was no minimum safety and security measures and no arrangement of hidden payroll (employee benefits and services or fringe benefit). And mostly they were lacking formal education and knowledge. Thus, workers were not in a position to bargain on any issue.

In this situation workers had no other option but to take political shelter. Political leaders on their part thought that if they could organize the large industrial labour force that would have strengthened their fight against the British. On this understanding political leaders came forward to form trade union. The trade union movement thus became an integral part of freedom movement. Though its main outlook was to use trade unions as an instrument to fight against the British, yet it did also some constructive work for labour movement in the beginning (Sharma, 1990 : 90). It was a necessity of the time, but this practice of taking political shelter gave a pattern of the trade union structure that could be characterised as the week organisation without the political support. This practice made them so weak that they could not conduct collective bargaining at the enterprise level without the political backing (Mondal, 1992: 8).

After a short while of the birth of Pakistan in 1947, political instability consequent upon the military rules hampered the development of a sound trade union structure in Pakistan (Mondal, 1992: 3). Division, disunity and disintegrity were a common phenomenon which was more aggravated by the leadership conflict. Moreover trade union activity was absolutely minimum in the areas of contemporary Pakistan. Major centers of trade union activity in British India, were located in Bombay, Calcutta, Madras and Kanpur etc. The territories then forming Pakistan, being predominantly agricultural, contained limited base for trade unionism. There were only 181 registered trade unions in Pakistan in 1948 and by December 1951 the number rose to 309 with membership of 3,93,137 (Mahmood, 1986).

After the war of liberation in 1971, the situation did not improve, but in fact deteriorated, in independent Bangladesh because of division and disintegration among workers as well as among the unions influenced by political parties (Bhuyan, 1991: 155-156). Due to a change in the economic policy of the country after liberation, 80% of the major industries and services including banks and insurance had been nationalised. So the role of trade unions became less important regarding wages and other benefits as the National Wage and Productivity Commission fixed it. Still the practice is on.

This chapter traces out the growth of trade union movement from 1947 to 1971 in different phases with political, social and economic changes that took place in Pakistan. It also focuses on the achievement of trade union movement for the working class in economic and social spheres of life.

Growth of the Movement During 1947 to 1957

We have noted earlier that industrial units were very few in number in Pakistan in comparison to Indian union. After partition and independence, Pakistan received 16 out of total 451 cotton mills, 9 out of 160 sugar mills, 5 out of 18 cement factories and none of 91 jute mills and 35 iron steel mills (Ahmed, 1978 : 37). As industries were very few in number, naturally the number of workers was also less. Just before partition, in August 1947, there were about 95,000 workers in the areas of Pakistan, of which 50,000 were in West Pakistan and 45,000 in East Pakistan (I.L.O., 1962 : 2). As the numbers of workers were small the trade union activity was very limited whatever the trade union activity was in Bengal (East Pakistan), it was the contribution of AITUC. It is to be recognised that the leftist labour leaders played a very important role in the wake of independence. Another notable fact is that responsible Muslim labour leaders never allowed communalism or raise its head in the labour movement (Ahmed, 1978 : 31).

Just after the birth of Pakistan 'Eastern Pakistan Trade Union Federation' (EPTUF) was formed on September 28, 1947. The government of Pakistan announced its first labour policy in 1948, which affirmed that the growth of genuine and healthy trade unions would be encouraged to promote collective bargaining with full appreciation of national economy. It was also stressed that trade unionism would be saved from exploitation and the work-force would be protected from victimisation for legitimate trade union activities. But nothing was done potential to reflect this policy either in Trade Union Act of 1926 or the Industrial Dispute Act, of 1947 (as adopted in Pakistan). The policy soon lost its impact as the industry developed fast and the employers grew powerful, the trade unions remained weak and financially insecure. Moreover, the policy fell short of achieving desired objectives because the policy was not backed by appropriate legislation (Mahmood, 1968 : 271). Within a short while of the formation of EPTUF conflict mounted up with the leftist. The final break up with the leftist came on March 16th, 1949 when the executive committee by a majority condemned the leftist dominated 'East Bengal Railroad Workers Union' for serving strike notice without prior approval of the federation. After the incident the communist members resigned from the federation. On 30th August, 1949 the EPTUF General Council expelled East Bengal Railroad Workers Union, Dacca District Textile Workers Union, Dacca Rickshaw Drivers Union, Barisal Bidi Workers' Union and Inland Steam Navigation Workers' Union from the federation (Ahmed, 1978 : 31-33). Thus from the very beginning division and disunity started in trade union arena which gave advantage to the management. Political and ideological conflict causes serious setback to the development of the trade union movement specifically in developing countries.

The first tripartite labour conference was held in Karachi on 8th and 9th February, 1949 with Hon'ble Labour Minister Jogendra Nath Mandal in the chair. Hon'ble Prime Minister Liakat Ali Khan opened the conference. Faiz Ahmed from East Pakistan attended the conference as a delegate where some union leaders from West Pakistan suggested to Faiz Ahmed that they would like to work together. Keeping in view of this suggestion a meeting of General council of EPTUF was called on 28th October, 1949 and was decided in the meeting that to organise a federation embracing unions in both parts of the country and the name of EPTUF was changed to All Pakistan Trade Union Federation (APTUF) (Ahmad, 1978 : 33).

Another Federation in the name of the Trade Union Federation, Pakistan (TUFP), Dacca was formed in January, 1949 with Nurul Huda as President, Kamruddin Ahmad as General Secretary and B.A. Siddique as Treasurer. By 1949 there were four major federations competing with each other. Later on APTUF was amalgamated with the Trade Union Federation, Pakistan (TUFP) on 30th April, 1950, after a long and protracted negotiation between the leaders of the two federations. According to agreement between the two federations Nurul Huda became the President of the APTUF and Kamruddin Ahmad, its Vice-President. Faiz Ahmad continued as the General Secretary and Aftab Ali, as Treasurer Abdul Awal and Zahur Ahmad Chowdhury were elected as Assistant Secretaries (Ahmad, 1978 : 33-34). However, we see the unification spirit of leaders in the early age of Pakistan.

As Dr. Malik was sworn in as the Central Labour Minister since then he began to pursue the federation leaders to form a National organisation. He contacted Mr. Khatib, the leader of Pakistan Federation of Labour in Karachi and Faiz Ahmad in Dhaka. The minister was successful to make them agree to form one national organisation. In the tripartite conference held in Karachi in 1950, Mr. Khatib proposed the leaders of East Bengal to unite to a federation of the two organisations and the name All Pakistan Federation of Labour was suggested. But East Bengal labour leaders specially Aftab Ali and Kamruddin Ahmad agreed upon to have a confederation (Ahmad, 1978). Last of all good sense prevailed, and in 1950 some of these federations merged together and formed a united platform by establishing 'The All Pakistan Confederation of Labour' (APCOL)[1] with two constituents units of East Pakistan Federation of Labour (EPFL) and the West Pakistan Federation of Labour (WPFL). This enable them to deal with the problems faced by the unions of a particular wing, the constituent units were given full autonomy (Rahman, 1969 : 39-40). Of the two constituent federations, the West Pakistan Federation of Labour was by far the more important and claimed two-third of the total membership. The APCOL was affiliated with the International Confederation of Free Trade Unions (D'Costa, 1963: 104).

The first convention of the national organisation was held with the labour leaders from East and West Pakistan on September 9th, 1950 in which 750 delegates represented about 320,000 organized workers (Quoted in Rahman, 1968 : 51).

A constitution of A.P.C.O.L. was drawn up on the basis of the draft prepared by Aftab Ali and an exclusive board consisting of thirty members was formed. The first office leaders were:

Portfolios	*Provinces*	*Names*
President	Karachi (Sind)	M.A. Khatib (Karachi)
Vice President	East Bengal	Aftab Ali (E.B.)
	West Pakistan	Umar Din (W.P.)
	Sind	Abdul Ghafoor (Sind)
General Secretary	East Bengal	Faiz Ahmed (E.B.)
Assistant Secretary	Karachi (Sind)	C.P. Dave (Karachi)
	East Bengal	M. Suleman (E.B.)

Source : Ahmed, 1978.

Out of 21 members of the first Executive Board 10 were from East Bengal; prominent among them were : Dr. A.M. Malik, Kamruddin Ahmed and Ghulam Murtoza (Ahmed, 1978: 35). Later on more federations were established like mushrooms.

As a result of personality clash and misunderstanding Nurul Huda along with A.R. Sunnamat left East Pakistan Federation of Labour and formed the Pakistan Labour Federation (PLF). Nurul Huda was elected its President, Nazir Mostafa, its General Secretary and A.R. Sunnamat, its Assistant Secretary. Due to this rift East Pakistan Railway Employees League (EPREL), Telegraph Association and Shops Association left East Pakistan Federation of Labour (EPFL). Pakistan Labour Federation continued with sincere effort of Sunnamat till 1956 when Pakistan Mazdoor Federation (East Pakistan Zone) was formed, Sunnamat as the President and K.A. Hye as its Secretary, Khaja Muhammad Hussain, President West Pakistan Mazdur Federation visited Dhaka and the two federations were amalgamated. Mazdur Federation used to be controlled by and led by the Tejgaon industrial area. There are all small industries but the unions ran on democratically. (Ahmed, 1978 : 35-36).

With the gradual development of industries trade union movement was also making headway. The following table 3.1 and 3.2 pertaining to June and December, 1955, show the satisfactory progress of trade union movement in East Pakistan.

Table 3.1 : Distribution of Unions and Number of Members in Eastern Pakistan on the Basis of Branch of Industries

Branch of Industries	*Number of unions*	*Percentage to total*	*Total number of members*	*Percentage to total*
Railways	5	2.7	58,544	38.6
Other transport	29	15.7	20,468	13.5
Cotton Textiles	17	9.2	11,874	7.8
Jute	23	12.4	18,921	12.5
Seamen	3	1.6	3,135	2.1
Municipal	5	2.7	774	0.5
Printing Presses	5	2.7	733	0.5
Dock	2	1.1	3,315	2.2
Port Trust	5	2.7	4,343	2.9
Engineering	6	3.2	801	0.5
Tea	3	1.6	8,850	5.8
Power Electricity	7	3.8	539	0.4
Commercial	16	8.6	4,753	3.1
Miscellaneous	59	31.9	14,552	9.5
All Industries	185	100	151,602	100

Source : Government of Pakistan, 1955.

From the table 3.1 above, we find that as a single industry jute occupies the highest number of unions. But as far as the number of members are concerned railway occupies the highest position. It is due to the fact that unions at plant levels are widely practiced in jute Mills. In railways, unions are mainly found at macro level.

The table 3.2 shows the classification of trade unions by number of members.

Table 3.2 : Distribution of Trade Unions in East Pakistan According to Number (as on June 30, 1955)

Membership	*No. of Union*		*No. of Members*	
	Actual No.	*Percentage to total*	*Actual No.*	*Percentage to total*
Below 50	23	12.4	711	0.5
50 and below 100	23	12.4	1,592	1.1
100 and below 300	69	37.3	12,349	8.1
300 and below 500	19	10.3	6,886	4.5
500 and below 1,000	29	15.7	18,272	12.1
1,000 and below 2,000	8	4.3	11,236	7.4
2,000 and below 5,000	9	4.9	25,161	16.6
5,000 and below 10,000	4	2.2	28,395	18.7
10,000 and below 20,000	-	-	-	-
20,000 and above	1	0.5	47,000	31.0
Total	185	100.00	151,602	100.00

Source : Government of East Pakistan, 1955.

From the above table we find that unions consisting of 100 and below 300 members were the highest in number at that time.

In order to show the rapid progress of trade union movement of Bangladesh (the then East Pakistan) within just six months' time we have furnished data in table 3.3 and 3.4 below pertaining to December, 1955.

From table 3.3 below, it is clear that jute occupies the highest number of unions as a single industry and railway had the highest number of members on 31st December, 1955. The increase of number of trade unions and the number of members are visible.

Table 3.4 below displays the number of unions according to the number of members as on 31st December 1955.

It is clear from the above table 3.4 that the highest number of unions had the membership of 100 and below 300. However, as far as actual total membership is concerned, unions having membership of 500 and below 1,000 groups occupied the top rank.

Table 3.3 : Distribution of Number of Unions and Number of Members in Eastern Pakistan on the Basis of Branch of Industries (as on 31 December, 1955)

Branch of Industries	*Number of Unions*	*Percentage to total*	*Total number of members*	*Percentage to total*
Railway	6	2.5	58,644	36.3
Other transport	39	16.4	22,379	13.9
Cotton Textiles	21	8.8	12,499	7.8
Jute	28	11.8	20,540	12.7
Seamen	3	1.3	3,135	1.9
Municipal	5	2.1	774	0.5
Printing Presses	5	2.1	733	0.5
Dock	2	0.8	3,315	2.1
Port trust	5	2.1	4,343	2.7
Engineering	7	2.9	842	0.5
Tea	3	1.3	8,850	5.5
Power Electricity	7	2.9	539	0.3
Commercial	21	8.8	6,133	3.8
Post and Telegraph	1	0.4	195	0.1
Miscellaneous	85	35.7	18,431	11.4
All Industries	238	99.9	161,352	100

Source : Government of East Pakistan, December, 1955.

From a perusal of the two tables pertaining to 30th June and two tables of 31st December 1955, the rapid progress of trade union movement of that time in East Pakistan is very much clear. Actual number of membership went up to 161,352 from 151,602 within six months.

After partition, democracy could not gain a solid base in Pakistan due to different reasons. From early fifties change of Government at quick succession led the country towards instability. Political instability and later on military rule prevented the development of a constructive trade union structure. There was mistrust between employer and the workers as the employers were mostly non-locals in East Pakistan (Mondal, 1992 : 193). The territory of Pakistan was

Table 3.4 : Distribution of Trade Unions According to Number (as on 31st December June, 1955)

Membership	*No. of Union*		*No. of Members*	
	Actual No.	*Percentage to total*	*Actual No.*	*Percentage to total*
Below 50	33	13.9	981	0.6
50 and below 100	31	13.0	2073	1.3
100 and below 300	96	40.3	16655	10.3
300 and below 500	19	8.0	6936	4.3
500 and below 1,000	37	15.5	22915	14.2
1,000 and below 2,000	8	3.4	11236	7.0
2,000 and below 5,000	9	3.8	25161	15.6
5,000 and below 10,000	4	1.7	28395	17.6
10,000 and below 20,000	–	–	–	–
20,000 and above	1	0.4	47,000	29.1
Total	238	100.00	161,352	100.00

Source : Eastern Pakistan Labour Journal, 31st December 1955.

infrastructurally backward than Indian Union in all respects. So the country had to make steady progress with different limitations. From economic point of view, the period 1951-58 may be called a period of industrial foundation and groundwork. During this period a number of Government and semi-Government institutions were established to help develop industrial and commercial undertakings. As a result, there was a gradual growth of the non-agricultural sector of the economy (Rahman, 1963 : 52).

From 1950 to 1958 a number of political ups and downs took place in the country. In 1954 election was held in East Pakistan. The cabinets were reshuffled and dismantled and martial law was imposed. These political troubles also reflected in the socio-economic life of the country. During these years due to conflict amongst the leaders, the APCOL failed to give a unified leadership to the trade union movement. Instead of acting "selfless and honest for improving the working and living conditions of the toiling masses," the leaders of the APCOL "remained busy in maintaining their position, power and leadership" (Rahman, 1969 : 40).

With the gradual industrial development trade union movement was also going on somehow satisfactorily, though there were division, disunity and personality clash in trade union arena. During the period 1951-54, the number of registered trade unions increased from 309 with a membership of 410,755, but the growth was hampered after 1954. The opposition party (the United Front) came to power in East Pakistan in 1954. But only after few months the ministry was dissolved and the province was placed under the administration of a Governor deputed by the centre. In the same year the communist party of Pakistan was banned and a large number of political and trade union leaders were arrested (Rahman, 1967 : 110). This was a serious blow to the trade union movement of Pakistan as the leftists were very much active in providing leadership to the overall labour movement. During the Pakistan regime, Government used to interfere in the trade union activities from time to time (Hussain *et al.*, 19: 146). In 1955 the number of trade union membership and the number of disputes declined to a considerable extent. After two years trade union membership showed an upward trend in 1957 and the figure rose to 384.127 in October 1958. But in the same month another political change took place — Martial Law was proclaimed under the leadership of Field Marshall Mohammad Ayub Khan (Rahman, 1969 : 110-111). All political and trade union activities came to a halt. So the natural growth of trade union movement had been threatened to a great extend.

Progress During 1958 to 1962 Under Martial Law

Martial law was imposed in October 7, 1958 and it was lifted on June 8, 1962. During Martial Law, all democratic instructions were put under restriction until further order. Like political parties trade union organisations also lost the right to organise meetings and to address anywhere. So after the declaration of

Martial Law trade union activities declined. The very act of military take over shattered the confidence of workers as to the gain from trade unions or other democratic institutions were not materialised and as a consequence immediately after the declaration of Martial Law trade union membership declined sharply (Rahman, 1968 : 53).

After the partition of British India in 1947, Pakistan inherited all the labour policies, legislations and regulations that were enforced in undivided India. Frequent political changes took place in Pakistan and Bangladesh, and each new regime that came in power declared one new labour policy containing similar programmes with minor differences (Khan, 1986: 37). First five year programme of action in the field of labour was drawn up in India during October 1946 and it was adopted in Pakistan in February, 1949 in the first labour conference, consisting of the representatives of the Government, employers and workers (Khan, 1986 : 87). But from 1949 to 1955 nothing remarkable had been achieved regarding the interest of the working class. Without doing anything substantial in last five years, the Government made a formal announcement of its labour policy in 1955. This policy also did not bring any good to the workers like the previous one. The only achievement was the Workmen's Compensation (Amendment) Act that was passed by the legislature in March 1957. This amendment was for increased rates of compensation for accidents. And another amendment for Industrial Dispute Acts, providing for speedy remedy to the aggrieved workers in the case of violation of the condition of service, and conciliation of apprehended disputes, were made in the middle of 1958 (Khan, 1986: 88).

In February 1959, the Pakistan Government announced a 'Revised Labour Policy'. The need for a healthy and vigorous trade union movement in the interest of industrial and social peace has been recognised (D'Costa, 1963 : 103). But in Martial Law period workers used to feel insecured to work freely. Though the Government also tried to take some legislative measures as a step towards the implementation of its declared labour policy. The Trade Union Act, 1926 was amended in 1960 and 1961 by the Trade Union (Amendment) Ordinance, 1960 (Ordinance No. XIV of 1960) and the Trade Unions (Amendment) Ordinance, 1961 (Ordinance No. XI of 1961). The former ordinance provided for the recognition of trade unions and came into force on August 1, 1961. The later ordinance provided for the exclusion of outsiders from the executive of the unions and came into force from March 16, 1961 (Shafi, 1961: 323). So this Act debarred any outsider from becoming an officer of trade union, unless he was full time servant of the union.

The following other ordinances were promulgated by the Martial Law authority before the imposition of a constitution by Ayub Khan in 1962.

1. The Industrial and Commercial Employment (standing orders) Ordinance, 1960.
2. The Working Journalists (condition and service) Ordinance, 1960.
3. The Minimum Wage Ordinance, 1961.
4. The Road Transport Workers Ordinance, 1962.
5. The Tea Plantation Ordinance, 1962.
6. The Merchant Shipping (Amendment) Ordinance, 1960.

Under the 1962 constitution, labours by and large become a provincial subject. The central labour laws were passed on to the Provincial Government. It is very interesting to note here that Pakistan Central Government was very much orthodox regarding the provincial autonomy and decentralisation of power. But the question why they segregated labour as a provincial subject. A conclusion may be drawn regarding the matter that it is troubled subject let it be in the hands of the provinces. Moreover, there is no chance of direct monetary gain and not the issue directly related to the security of the country. Following the transfer of this subject to the province, the Government of East Pakistan enacted several Acts by repealing the central Acts and Ordinances. The most important among the enactments by provincial legislatures were as follows :

1. The East Pakistan Trade Union Act, 1965;
2. The East Pakistan Labour Disputes Act, 1965;
3. The East Pakistan Employment of Labour (Standing Orders) Act, 1965;
4. The East Pakistan Factories Act, 1965;
5. The East Pakistan Shops and Establishments Act, 1965; and
6. The East Pakistan Water Transport (Regulation of Employment) Act, 1965. (Ahmed, 1978 : 70-72).

These Acts gave a solid trade union; recognition and the employer had hardly any say against the trade union recognition. Accordingly, Labour Dispute Act, 1965 provided for collective bargaining but for all practical purposes, the use of strike was made impossible. Strikes were banned in public utility services. Employment of Labour (standing orders) Act, 1965 the most unfavourable provision was the termination of employment without showing any reason which was against the very spirit of collective bargaining. The Factories Act, 1965 incorporated some additional provisions with regard to leave of the workers (Khan, 1986 : 90-91).

The following table 3.5 will provide a picture of Registered Trade Unions with membership in Pakistan till 1962.

Table 3.5 : Registered Trade Unions with Membership in Pakistan

Period	*Number of Unions*	*Total Membership*
1950	251	Not Available
1951	309	393.137
1952	352	394.923
1953	394	424.563
1954	382	410.755
1955	474	325.610
1956	524	316.317
1957	611	366.317
1958	621	357.064
1959	618	347.522
1960	708	350.604
1961	723	398.723
1962	789	417.248

Source : Government of Pakistan (1966 : 582).

Period from 1962 to 1971 : Labour Movement Under Restricted Democracy

Martial Law was lifted on June 8th, 1962. In the June-July session of the national assembly in the same year, the ban on the formation of parties was lifted and democratic and constitutional form of Government was restored and the right to form and join associations was assured by the Constitution (Khan, 1980 : 294-295). In addition to the All Pakistan Federation of Labour (APCOL) the Pakistan National Federation of Trade Unions (PNFTU) was formed in September 1962, and has been affiliated with the ICFTU with effect from 1st July 1966 (Rahman, 1967 : 112). There were 965 registered trade unions with a membership of 512225 in December, 1965 (GOP, 1966 : 582), whereas in December, 1961 there were only 723 registered trade unions with 398,723 members (GOP 1966 : 582). This progress was due to restoration of constitutional democracy.

Though communist party of Pakistan was banned and a number of political and trade union leaders were arrested in 1954, the communists were not sitting idle. They began to concentrate their activities in plantation, cement factories, Fenchuganj fertiliser factory and jute mills in Chittagong division, cotton textiles specially in Tongi, cigarette factories, soap factories, sugar mills, biscuit factories, railway and Biri Sramik in North Bengal (Ahmed, 1978 : 46). They used to work with caution or underground as their party was banned. During communal riots in 1963, which began in Khulna and later on spread to Narayanganj, Demra, Tongi and Narshingdi, leftist decided to resist such conspiracy to break the unity among the workers. Trade union leaders like Mahbubul Haque, Farid Ahmed, Kazi Muhiuddin M. S. Haque, Abdul Jabbar (Khulna) were elected to convene a preparatory committee meeting for calling an all East Pakistan workers conference in August, 1964. But the conference was postponed on the request of Chatkal Srammik Federation and Mazdoor Federation. The conference was ultimately held on October 23rd and 24th, 1964, in which delegates of 124 unions attended (Ahmed, 1978: 46).

East Pakistan Federation of Labour (Branch of all Pakistan Confederation of Labour) split up into two factions in 1963—one led by late Mr. Aftab Ali and the other by the late Mr. Faiz Ahmed (Hafiz, 1980). So disunity and disintegrity was the characteristics of the trade union movement in Pakistan. And as we have mentioned earlier, leadership conflict was the reason behind these divisions. Another organisation was formed in the name of "Purba Pakistan Sramik Parishad", but most of the leaders lost interest after Indo-Pakistan war in September 1965, after which defense of Pakistan rule was promulgated and the organisation survived as a federation of the leftist with the name of Purba Pakistan Sramik Federation with Mohammad Toha as the President and Sirajul Hussain Khan as the General Secretary (Ahmed, 1978 : 46). The growth of trade union movement during 1951-1966 has been shown in the table 3.6 below:

Table 3.6 : Growth of Trade Unionism in Pakistan

Year	*Trade Unions*	*Membership*
1951	209	3.9 lacs
1961	708	3.5 lacs
1964	898	4.0 lacs
1966 (May)	988	5.2 lacs

Source : Eastern Worker, 1968, Vol. VIII.

It is clear from table 3.6 that in 1951 the number of trade unions was 209 and the membership was 3.9 lacs. In 1961, the number of trade unions rose to 708, but membership did not increase rather went down from 3.9 lacs to 3.5 lacs. So multiplicity of trade unions had been started much earlier. In the succeeding years number of membership increased but not in proportion to number of trade unions. Multiplicity of trade union is the net result of disunity, disintegrity among the labour leaders and workers and outside leaders' influence.

In 1964, jute workers' historic strike continued for about eight weeks. The strike was organised by jute mills workers union under the leadership of Aftab Ali with a new charter of demand submitted on July 1, 1964 following the expiry of the agreement on June 30th, 1964. Other five unions of Dacca, Chittagong were claiming their allegiance simultaneously with Purba Pakistan Sramik Federation, a rival federation of Pakistan jute mills workers union issued notices on their respective employers on the same date expressing their intention to go on strike from July 15, 1964. The strike actually commenced on July 31. Later on an interim agreement was signed on August 5, 1964 and the strike was called off under the direct intervention of the Governor of East Pakistan. Under the leadership of Abdul Mannan and Maulana Syudur Rahman Purba Pakistan Sramik Federation declared strike from the October 20 and continued till a settlement was reached in December 7, 1964 with other benefits. The minimum wages was raised from Rs. 65 to Rs. 81 (Ahmad, 1987: 47-48). Jute was the biggest sector at that time and the trade union movement was also strong in it.

It is revealed from the above table 3.6 that from 1965-66 onwards number of workers was rising gradually but in 1971-72 number of workers declined to a considerable amount. The reason might be that after liberation mills and factories could not open due to different reasons. But the Government was careful as it was the biggest source of foreign earnings and at the same time jute was the most important employment generating industry.

Table 3.7 : Number of Workers in Jute Mills

Year	*Number of Mills*	*Number of workers (000) Permanent*
1965-66	23	80
1966-67	29	100
1967-68	34	110
1968-69	47	150
1969-70	55	175
1970-71	68	170
1971-72	68	153

Source : Government of Bangladesh, 1980, p. 32.

After lifting of Martial Law, trade union movement was in full swing, though due to leadership conflicts there was some division and disunity amongst the workers. Employer had always the tendency to exploit the workers, and the Government always supports the employers specially in the developing countries like Pakistan. Some other organisations came up in sixties. Another federation named "Jatiyo Sramik Federation" was formed in 1966 with Mr. Abdul Bashar as President. "Purba Pakistan Sramik Federation" was formed in 1967 with Mr. Tooha as President (Hafiz, 1980).

Not only federations, number of workers and number of unions were also increasing gradually with the gradual industrial development, we can have a clear picture of 1966 from the table 3.8.

Table 3.8 : Classification of Trade Unions According to the Number of Workers as on 31st March, 1966

Membership	*No. of Union*		*No. of Members*	
	Actual No.	*Percentage to total*	*Actual No.*	*Percentage to total*
Below 50	54	14.83	1,763	0.53
50 and below 100	55	15.11	3,947	1.18
100 and below 300	113	31.04	21,441	6.41
300 and below 500	53	14.56	21,292	6.38
500 and below 1,000	33	9.06	24,098	7.22
1,000 and below 2,000	35	9.62	48,507	14.52
2,000 and below 5,000	14	3.85	39,718	11.89
5,000 and below 10,000	4	1.10	29,720	8.90
10,000 and below 20,000	1	0.27	19,431	5.82
20,000 and above	2	0.55	124,068	37.15
Total	364	100.00	333,985	100.00

Source : Government of Pakistan, March, 1966.

From table 3.8 below it is clear that small plant level unions having membership of 100-300 workers were much popular. Small sized unions were good in number because plant level unions were always in practice in this country. As size of the plants were small, maximum number of unions were also small from the membership point of view.

From the following table 3.9 we can assess the growth and development of trade unions within a year or so in the then East Pakistan.

Table 3.9 : Classification of Trade Unions According to the Number of Workers as on 31st December 1966

Membership	*No. of Union*		*No. of Members*	
	Actual No.	*Percentage to total*	*Actual No.*	*Percentage to total*
Below 50	54	14.40	1,599	0.50
50 and below 100	53	14.13	3,811	1.12
100 and below 300	119	31.73	22,638	6.62
300 and below 500	58	15.47	23,048	6.74
500 and below 1,000	34	9.07	24,798	7.26
1,000 and below 2,000	34	9.07	47,265	13.83
2,000 and below 5,000	16	4.27	45,265	13.24
5,000 and below 10,000	4	1.07	29,720	8.70
10,000 and below 20,000	1	0.27	19,431	5.69
20,000 and above	2	0.53	124,068	36.30
Total	364	100.00	341,734	100.00

Source : Government of Pakistan, 1966.

Awami League entered into the trade union arena very late. Leftist had a strong hold and Muslim League had little influence in the labour front. Before liberation of Bangladesh, the Awami League wanted to have its own trade unions in major industrial centers like Tejgaon, Tongi, Khulna, Chittagong and in other areas of the country. Sheikh Mujib created Jatiyo Sramik League in 1969 to have control over trade unions (Bhuyan, 1991 : 152), Jatiyo Sramik League was formed with Mr. Nurul Haque as President and Abdul Mannan as General Secretary. At that time trade union Kendro was also formed with Mr. Harunur Rashid Chowdhury as President and Mr. Saifuddin Manik as General Secretary (Hafiz, 1980).

Everything was going on as usual, but again the country faced another Martial Law. President Ayub Khan resigned and handed over the power to the Commander-in-Chief of the Army, General Yahya Khan on March 25, 1969. The whole of Pakistan was again declared to be under Martial Law by General Yahya Khan (Kha, 1980 : 295). But this Martial Law did not continue for a long time. On December 21, 1969 a political activities regulation No. 60 of 1969 was issued and the following day the ban on political activities was lifted by another regulation and there by on January 1, 1970 normal political activity was permitted again within the standards of conduct laid down in Martial Law regulation 60 (Khan, 1980 : 295).

The whole of Pakistan was passing through serious mass agitation combined with industrial unrest throughout the second half of the sixties, which reached its peak in 1969. Labour front was also very much annoyed regarding the

Table 3.10 : Classification of Trade Unions According to the Number of Workers as on 30th June, 1968

Membership	*No. of Union*		*No. of Members*	
	Actual No.	*Percentage to total*	*Actual No.*	*Percentage to total*
Below 50	54	13.46	1,573	0.48
50 and below 100	62	15.46	4,351	1.26
100 and below 300	127	31.67	24,245	7.03
300 and below 500	64	15.96	25,071	7.26
500 and below 1,000	36	8.98	26,592	7.71
1,000 and below 2,000	36	8.98	50,033	14.51
2,000 and below 5,000	16	4.00	45,252	13.11
5,000 and below 10,000	3	0.74	24,315	7.05
10,000 and below 20,000	1	0.25	19,431	5.63
20,000 and above	2	0.50	124,068	35.96
Total	364	100.00	345,031	100.00

Source : Ahmad Kamruddin, 1978: Appendix VI.

Table 3.11 : Statement Showing the Distribution of Workers in East Bengal by Industrial Groups up to 1968

Sl. No.	*Name of Industry*	*Workers employed in East Bengal*
1.	Jute Manufacturing (Jute Mills)	79,712
2.	Cotton Textiles, Ginning and Baling	33,627
3.	Jute Processing and Baling (Jute Press)	4,378
4.	Sugar Mills	13,464
5.	Rubber Plantation	N.A.
6.	Plantation (Tea)	1,01,248
7.	Match Industry	10,088
8.	Shoe Industry	249
9.	Leather	620
10.	Silk & Rayon (including textile)	95
11.	Iron & Steel Re-rolling Mills	972
12.	Cement, Lime and Potteries	778
13.	Shipyard & Ship buildings	1,412
14.	Fertilizer Factories	–
15.	Chemical (including Pharmaceutical & Cosmetics)	5,003
16.	Iron Foundries	1,839
17.	Paper and Newsprint	4,993
18.	Glass	2,114
19.	Ceramics	390
20.	Food (Biscuit etc.)	1213
21.	News Papers	817
22.	Tobacco	3,576
23.	Transport (Rail)	76,474
24.	Transport (Road)	22,51
25.	Transport (Inland Water)	N.A.
26.	Mines and Quarries	473
27.	Ports and Docks	15,000
	Total	3,82,346

Source : Ahmad Kamruddin, 1978 : Appendix IX.

Table 3.12 : Classification of Trade Unions by Industry as on 30th June, 1968

Branch of Industry	*No. of Unions*	*Percentage to total*	*No. of Membership*	*Percentage to total*
Railway	4	1.00	48,106	13.94
Other Transport	36	8.98	26,027	7.54
Cotton Textiles	47	11.72	30,080	8.72
Jute Textiles	37	9.22	35,627	10.33
Tobacco	24	5.99	8,124	2.35
Municipal	7	1.75	1,893	0.55
Printing Press	19	4.74	3,112	0.90
Sugar	10	2.49	17,169	4.98
Dock	1	0.25	8,758	2.54
Paper Mills	3	0.75	4,110	1.19
Port Trust	6	1.49	8,563	2.48
Match	9	2.24	4,155	1.20
Engineering	19	4.74	5,841	1.69
Tea	3	0.75	1,00,576	29.16
Power Electricity	9	2.24	3,641	1.06
Glasses	7	1.75	2,661	0.77
Commercial	51	12.72	8,605	2.49
Post & Telegraph	1	0.25	9	0.00
Miscellaneous	106	26.43	23,827	6.91
Total	401	100.00	3,45,0.31	100.00

Source : Ahmad Kamruddin, 1978 : 78, Appendix VII.

Government politics and they used to take part with the anti-Government agitations. Labour disputes in East Pakistan almost became a regular feature at that time (see Appendix-I). In this situation Martial Law was imposed. The new Martial Law Government, headed by Yahya Khan, showed a sympathetic attitude towards workers and declared a new labour policy in July, 1969 which frankly accepted that workers had been deprived of their genuine rights in the past. It is claimed that the policy was far more accommodative than those declared previously and even subsequently after the independence of Bangladesh (Khan, 1986 : 91).

A new ordinance, called Industrial Ordinance, 1969 was promulgated which may be regarded as an important landmark in the history of labour. The Ordinance concerns the formation of trade unions and settlement of disputes between employer and workmen. Previously there were two separate laws on each subject in each province. These consolidated into this ordinance which was applicable throughout Pakistan - East Pakistan as well as West Pakistan. The Trade Union Ordinance and the Industrial Disputes Ordinance applicable in each province stood repealed (Shafi, 1969 : 3-4).

The Minimum Wages (Fixation) Ordinance, 1969 was also promulgated for implementing the objectives cherished in the new policy. It fixed the rates of minimum wages for all industrial establishments, public or private, having 50 or more workers. So, at the fag end of the Pakistan period, all workers, both in public and private sectors, were allowed freedom of association and right of collective bargaining, including the right to strike. The minimum wages fixed were applicable irrespective of private and public workers of essential and service sectors like posts, telegraph and telephone, electricity generation, railways etc. were allowed the right to strike (Khan, 1986 : 91-92).

Gherao movement of 1969 deserves discussion in brief as an important event in the trade union movement of Pakistan. The struggle against the dictatorial regime of Mohammad Ayub Khan started in the towns and cities in West Pakistan. Lawyers and students of the university took the leadership. Political leaders took the advantage of the discontents among the students and workers and mobilized them against the regime. The people of East Pakistan did not take the movement seriously in the beginning except the National Awami Party (Moscow group). The students—The East Pakistan Students League, The East Pakistan Students Union (both Moscow and Peaking groups) protested against the police action in the West Pakistan on November 23, 1968. Later on there was a public demonstration starting from Baitul Mokarram in Dhaka on November 24, by the NAP (Moscow). From the end of November, 1968 Maulana Bhasani, President of NAP (Peaking) called upon the rural poverty stricken peasantry to *'Gherao'* (surround) the houses of the corrupt development officers and Tahsil offices. In many places '*Gherao*' stated from December. It is worth mentioning that the peasant revolt in Naxalbari in West Bengal (India) in 1966 had its impact on the politically conscious peasant leaders of East Pakistan (Ahmed, 1978: 78).

The *Gherao* movement started in rural areas by Maulana Bhashani had its impact on the proletariat sections of industrial areas and the '*Gherao*' by the employers became a regular feature from first week of March 1969. But the *Gheraos* were not only confined within the workers of the factories and offices, rather the movement spread among the students too. Students were also involved in the movement inspired by political parties—both National Awami Parties (Moscow and Peaking group) and Awami League. This *Gherao* movement must have far reaching impact on society though apparently it might be thought that everything was settled with the fall of Ayub Khan on 25th March, 1969 (Ahmed, 1978 : 85-86).

The people and workers in 1969 went through 'Gherao' and mass movement. As a result there was a change of Government (Hafiz, 1980). But another military bureaucrat appeared on the scene. Throughout the Pakistan period some Rushputins of the cantonment used to control everything behind the scene, sometimes they used to appear on the political dais. New Martial Law Authority announced that no retrenchment/ termination of services of workers should be done without the prior approval of the Martial Law Authorities on 20th May, 1969 (Ahmed, 1978 : 93).

The Government took some positive steps for the workers. This Government arranged National Election to transfer power to the people's representative. The Government seemed to be better than other military Government up to the election. But after the General Election of 1970, the situation deteriorated as the power was not handed over to the representatives of the majority elected members (Hafiz, 1980). When Bengalees bagged the majority seats of the national assembly then Rushputins of the cantonment and Mr. Z.A. Bhutto (Peoples

Party) arranged a nasty and nepherous drama to foil the transfer of power. The Bengalees were forced to take arms to protest against the crackdown and mass killing started by Pakistani army. Actually Pakistani administrator thrust the war of liberation upon Bangladeshis.

Workers, students, and army personnel, EPR and other professionals and people from all walks of life and irrespective of sexes did participate in the war of liberation. The Peoples' Republic of Bangladesh emerged as an independent sovereign state in the World Map on December 16, 1971, with the help of India and other countries.

Summary

In this chapter, we have tried to depict the organizational development of trade unions and their movement with the gradual industrial development of a newly born country, which was far backward industrially and economically as a whole than the Indian union. As the country was industrially poor, naturally trade union activity was absolutely minimum.

Factory system set out its journey just in the middle of 19th century with the erection of a textile mill in Bombay in 1851 and a jute mill in Calcutta in 1854. After partition of India Pakistan got very few industrial units in comparison to India. After partition Pakistan tried to concentrate to develop industrially but her internal problems did not allow doing so. After a short while of the birth of Pakistan in 1947 poor economic infrastructure, political instability, continuous military rule hampered industrial development and development of a sound trade union structure as well. Despite all the limitations organisational development took place and trade union movement made its headway.

Division, disunity, personality clash, ideological conflict and outsiders' dominance hampered the growth of trade union movement of Pakistan. The scenario of labour movement of this sub-continent was almost same. However, humble effort has been made to trace out the growth and development of trade union movement from 1947 to 1971 in different plases with political, social and economic changes that took place in Pakistan during the period.

REFERENCES

Ahmad, Kamruddin, (1978), *Labour Movement in Bangladesh,* Inside Library, Dhaka.

Akkas, Md. Ali and Hossain Ferdous, (1998), Characteristics and Weaknesses of Trade Unionism: Bangladesh Perspective, *Social Science Review,* Vol. XV, No. 2, Department of Sociology, University of Dhaka.

Bhuyan, M. Sayefullah, (1991), The Trade Union Politics in Bangladesh: A Study of Textile Workers Union, *Dhaka University Studies,* Part-C, Vol. 12 (1).

Cole, G.D.H., (1962), *A Short History of British Working Class,* Allen Unwin, London.

D'Costa, Ronald, (1963), *The Role of Trade Union in Developing Countries: A Study on India, Pakistan and Ceylon,* Havaux, Nivelles.

East Pakistan Labour Journal, 1966, Vol. XIX, No. 4.

East Pakistan Labour Review, Vol. XIX, No. 1, March, 1966.

Eastern Pakistan Labour Journal, December, 1955.

Eastern Pakistan Labour Journal, June 1955.

Eastern Worker, (1968) : Vol. VIII, ed. M. Shafi, Bureau of Labour Publication, Karachi, May 3.

Government of Pakistan (1960), *Trade Union Amendment) Ordinance.*

Government of Pakistan, (1961), *Trade Union (Amendment) Ordinance.*

Government of Bangladesh, (1980), *Statistical Year Book of Bangladesh.*

Government of Pakistan (1969), *Industrial Relations Ordinance,* 1969, with commentary M. Shafi, Nov. 1969, Bureau of Labour Publications, Karachi.

Hafiz, S.M.A., (1980), *The Bangladesh Observer.*

Hussain, Ismail M., Hye, Syed A. and Ali, A.M., (1996), Structural Adjustment Policies and Labour Market in Bangladesh, ed., Mustafa K. Mujeri, *Centre on Integrated Rural Development for Asia and the Pacific,* (CIRDAP) Dhaka.

Khan, Hamiduddin, (1980), *The Fundamental Right to Freedom of Association in the Indo-Pak-Bangladesh Subcontinent, University of Dhaka,* Bangladesh.

Khan, Md. Abbas Ali, (1986), Government Policies Towards Labour in Bangladesh : A Historical Analysis, *The Dhaka University Studies*, Part-C, 7 (2): 1986.

Labour Code of Pakistan (Central).

Mahmood, M., (1968), Trade Unionism in the Last Ten Years, *Eastern Workers*, September, 1968, Vol. VIII, No. 9.

Malik, A.M., (1954), *Labour Problems and Policy in Pakistan*, Pakistan Labour Publications, Karachi.

Mondal, Abudl Hye, (1992), Trade Unionism, Wages and Labour Productivity in the Manufacturing Sector of Bangladesh, Research Report No. 133, Bangladesh Institute of Development Studies (BIDS), Dhaka.

Morning News, Dacca, 8th June, 1962.

Mukherjee, R., (1951), *The Indian Working Class*, Hind Kitab Ltd., Bombay.

Pakistan Labour Gazette, Vol. XIV, No. 3, July-September, 1966.

Quoted in Khan M.M. and Ahmad M., (1980), *Participative Management in Industries*, Centre for Administrative Studies, Dhaka.

Rahman, M.A., (1967), Laour Union and Political Organisation, University of London, *Institute of Commonwealth Studies*, January-May, 1967, No. 3.

Rahman, M.A., (1968), *Trade Unions in Pakistan: An Evolution of Their Problems and Development*, unpublished dissertation of Degree of Master of Science (Social Science) of the University of Edinburgh.

Rahman, M.A., (1969), Trade Union Management and Leadership in Pakistan, *Eastern Worker*, Vol. IX, NO. 6 & 7, Karachi.

Rahman M.A., (1969), Structure of Trade Unions in Pakistan, *The Pakistan Economic Review*, Vol. 1, No. 1, Winter, 1969.

Report on Workers Education in Pakistan (1962), *ILO Report*, Geneva.

Sharma, Ganga Shahi (1990), *Trade Union Freedoms in India*, Deep and Deep Publications, New Delhi.

Siddique, Shafiq Ahmad, (1989), The Imperial Interest and Development of the Industrial Relations System in Colonial India, the Dhaka University Studies, Part-C, Vol. 10 (2), 1989.

Weatherford, W.D., (1962), 'Pakistan', in Galenson, W., (ed.) *Labour in Developing Economics*, University of California Press.

APPENDIX–I

Labour Disputes in East Pakistan since 1947

Period to which the Figure relates	*Nomber of disputes resulting in stoppage of works*	*Number of workers involved*	*Nomber of man-days lost*
1	2	3	4
1947 (for the period from 15th August	26	12,091	18,963
to 31st December)	48	28.224	112,163
1948	54	29,839	82,839
1949	20	7,972	40,381
1950	23	11,628	35,792
1951	55	21,174	75,624
1952	55	34,006	116,080
1953	31	19,738	90,556
1954	25	16,221	38,940
1955	64	82,261	273,632
1956	71	115,249	693,167
1957	51	45,903	152,351
1958	7	22,488	35,482
1959	12	6,086	17, 947
1960	10	4,464	5,486
1961	31	16,949	85,248
1962	54	102,198	938,093
1963	72	158,614	3,787,357
1964	55	63,707	236,805
1965	60	78,278	241,100
1966	48	66,391	605,222
1967			
1968	3	1,322	2,767
January ...	1	134	938
February ...	3	5,346	13,798
March ...	3	7,212	42,207
April ...	3	1,696	3,537
May ...	4	30,896	37,303
June ...			

Source : Ahmad, Kamruddin, 1978 : xxviii.

Chapter - 4

GROWTH OF TRADE UNION MOVEMENT IN INDEPENDENT BANGLADESH

Introduction

On December 16, 1971 Bangladesh emerged as an independent nation on the world map. The people fought for the country with high hopes and aspirations. But the country had a war-ravaged economy. Industrial workers and agricultural labourers were displaced; some participated in the war of liberation. So, it took some time to settle everything. By this time Peoples Republic of Bangladesh as a state policy nationalized 80 percent of the major industries and services including banks and insurance. The decision of change over from one economic system to another was rather a bit early while the country just got liberation. However, there was a massive change in the economic system, it might be regarded as a forward step towards socialistic pattern.

Nationalization was almost a new concept to the workers and management people too. Corrupt politicians made the workers and labour leaders understand that after industries being nationalised, their services being nationalised, the factory was their own property. This is in fact, the basic theme of socialism where all are equal and everybody has equal access to all the existing resources available. But they interpreted the very logic of socialism and misused the assets of nationalized mills and factories to satisfy their self-interest. During this long period of 28 years Bangladesh experienced two Martial Laws, one in 1975 and the other in 1981, and during the Martial Laws the organized movement of workers in industries including jute was disrupted time and again.

In this changing circumstances how the workers, trade union leaders and management behaved, will be analysed in the present chapter.

Trade Union Movement from 1972 to 1981

As started earlier after liberation Government of the Peoples Republic of Bangladesh as a state policy nationalized about 80% of the major industries and services including banks and insurance. So there was a massive change in the economic system which was supposed to lead the country towards socialistic pattern. Immediately after liberation, the Prime Minister (Sheikh Mujibur Rahman, who led the liberation movement was the first Prime Minister of Bangladesh) had a press statement on February 9, 1972 which read as under—

> "I assure our workers that the basic goal of the socialist economy, which we are committed to achieve, will be securing the rights of workers and ensuring their welfare. A plan is being prepared where by measures of nationalization would be combined with new arrangements to ensure workers participation in the management of industries" (Quoted in Khan, 1980 : 56).

But the workers illogically demanded higher wage and fringe benefits just after liberation. The Government reacted positively. The Government constituted the Industrial Workers Wage Commission immediately on the demand of the workers. This Commission submitted the report to the Government with certain recommendations. The Government accepted the report. In order to implement these recommendations Government promulgated a law known as State Owned Manufacturing Industries Workers (Terms and Condition of Services)

Ordinance 1973. But workers did not take the Ordinance easily as they had lost their basic right of collective bargaining. The report of the Pay Commission for Government, autonomous and semi-autonomous employees was also published in 1973.

So wages and other benefits are fixed by the National Pay Commission (for Government and public sector employees) and National Wages and Productivity Commission (for industrial workers in the public sector) and Collective Bargaining Agent (CBA) has nothing to do in this regard (Akkas and Hussain, 1998 : 151). The main issue of bargaining is wage and other benefits. But the wage issue had been ceased by wage commission as a result of nationalisation. Leadership in plant levels and even federation levels had almost very little to do in financial matter. So they did not take the new economic system from the core of their hearts as it crippled their power.

Employers of different organisations also did not accept the scale with satisfaction. So large-scale agitation and conflict came up. Several strikes recorded in Jute and cotton industries that caused considerable loss of production (Hafiz, 1980). Actually the enhancement of wage and salary due to new wage and pay commission could not cope with the rate of inflation. Moreover, management workers' co-operation was absolutely minimum at that time. Workers were going to be unruly due to economic recession after liberation and they could not follow the meaning of nationalisation and socialism properly. In their perception, socialism meant that every employee is equal and there would be no difference between a general manager and a labour in respect of remuneration and status. They thought that there would be no hierarchy. And nationalisation means industry has become its own and their service has also been secured. So they used to defy hierarchical position and instruction issued by the superior authority. Moreover, executives had the paternalistic attitude and they were feeling uneasy to adjust with this new order. Thus, labour management cooperation came to a lowest extent. But to increase productivity and quality of work life labour management cooperation is essential. Besides that it is of much importance to establish a cooperative relationship between management and labour in order to succeed in the competitive business world (Akkas, 1998 : 118).

After the liberation of Bangladesh, the new Government adopted the Indusrial Relations Ordinance (IRO), 1969 (partially amended in 1970) with a view to regulating labour management relations along with some other laws as Pakistan Government adopted all the British India Laws after 1947 (Alam, 1983 : 44). First Industrial Policy was declared in 1972, which recommended for reducing the activities of the trade unions to the status of a welfare organisation. Right to strike in the nationalised industries was prohibited for six months. In the year 1973 the right to strike and lockout, as guaranteed in the year 1969 Ordinance was obliterated (Sayefullah, 1986 : 163). During the first Martial Law, Government of Bangladesh promulgated the industrial Relations (Regulation) Ordinance, 1975 (IR (R) O, S 1975) as an addition to the IRO, 1969. The Government amended the IRO, 1969 in July 1977 and the new ordinance was named as the Industrial Relations (Amendment) Ordinance, 1977 (IR(A)O, 1977). Further amendments came in 1980. IRO, 1969 was amended in July, 1980 and the new one was named as the Industrial Relations (Amendment) Act, 1980 (IR(A) A, 1980). This Act repealed IR(R)O, 1975 and incorporated some important features of the first Labour Policy declared in March 1980. After the IR(A)A, 1980 a new law, *i.e.*, the Industrial Relations (Regulation) Ordinance, 1982 was promulgated by the Second Martial Law Government (Alam, 1983 : 44).

Another amendment took place in 1985. This will be treated as Industrial Relations (Amendment) Ordinance 1985. Article 7 (a) of 23 Ordinance of 1969 had been amended where eligibility of being elected members or officer bearers of any trade union have been amended. In addition to that, article 10, 11, 47 and 47 (A) and 47 (B) of the said Ordinance of 1969 have been amended. All these amendments were related to cancellation of trade union registration, appeal organize cancellation, regarding the transfer of office bearers of trade unions, etc. (Industrial Relations (amendment) Ordinance, 1985), Ordinance No. 15 of 1985), Govt. of Bangladesh.

After liberation the Government was urged by Jatiyo Sramik League (JSL) to ban those labour unions which had co-operated with Pakistan occupation army directly or indirectly and take punitive measures against the leaders of those labour unions. And those leaders should not have any change in any ensuing tripartite committees. The then labour Minister Mr. Zahur Ahmad Chowdhury held the same view. Faiz Ahmad group was amalgamated with JSL for existence but Aftab Ali group kept its identity separate by electing Dinen Sen as President and expelling their president Mr. Saifur Rahman and their leading member Vilayet Hussain (Ahmad, 1978 : 101). Country needed national unity badly to rebuild the war-ravaged economy at that time. So JSL could not have brought those issues at this juncture.

By this time industrial unrest was cropping up. There was a *gherao* in the Pak Jute Bailing Company, the Kohinoor and Adamjee groups of industrial areas where workers compelled the management to surrender to their demands (Ahmad, 1978: 101-102).

Jute industry was the backbone of Bangladesh economy. Jute is used to contribute highest source of our foreign exchange. It is the biggest employer in the industrial sector of the country. Unlike any other manufacturing industry in the country, the Jute industry has to sell more than 90 percent of its produce in the international market in the face of severe competition from synthetics and other jute goods exporting countries like India, China, Thailand, Nepal etc. In spite of so many odds, Bangladesh occupies a key position in the world Jute goods export market (Bangladesh Economic Survey, 1984-85 : 44). But this sort of chaotic condition as noted above was a threat to the industry and a newly liberated country.

From the table 4.1 we can assess the position of jute manufacturing industry up to 1978-79. Number of permanent workers began to decrease after liberation due to unstable economic and political situation. Again it began to increase gradually from 1976-77. But Jute Mills were running in full swing.

Table 4.1 : Profile of Jute Manufacturing Industry in Bangladesh (1965-66 to 1978-79)

Year	*No. of Mills*	*No. of looms installed (a)*	*No. of looms (operating) (a)*	*No. of spindles worked (.000') (a)*	*No. of workers (permanent) (.000') (a)*
1965-66	23	13330	11868	178	80
1966-67	29	14603	13073	201	100
1967-68	34	18255	16927	258	110
1968-69	47	20573	18358	293	150
1969-70	55	21453	20631	302	175
1970-71	68	23781	18137	290	170
1971-72	68	23836	15188	250	153
1972-73	73	24184	19933	300	162
1973-74	73	23709	19334	N.A.	150
1974-75	77	24651	17565	N.A.	142
1975-76	77	25679	20271	N.A.	131
1976-77	77	25679	20261	N.A.	132
1977-78	77	25679	22168	346	135
1978-79	77	26034	20493	N.A.	142

Source : Government of Bangladesh, 1980.

The table 4.2 represents countries production of jute goods in the country in comparison to export, internal consumption and stock. Total production and internal consumption began to decrease after liberation due to instability prevailing all around. From 1976-77 it began to regain its position gradually.

Table 4.2 : Production of Jute Goods Compared with the Trends in Export, Consumption and Stock

Year	*Total Production*	*Export Overseas*	*Internal Consumption*	*Stock at*	
				Beginning of the season	*End of the season*
1955-56	130	85	40	13	11
1956-57	145	81	57	14	21
1957-58	158	104	57	18	15
1958-59	203	131	55	14	25
1959-60	256	197	51	24	32
1960-61	249	203	48	31	29
1961-62	271	216	57	28	26
1962-63	298	219	65	24	38
1963-64	331	250	69	35	47
1964-65	289	222	69	47	49
1965-66	410	316	71	49	71
1966-67	404	347	68	71	61
1967-68	509	422	74	61	73
1968-69	507	452	66	73	62
1969-70	561	496	63	62	63
1970-71	440	377	56	63	70
1971-72	315	238	45	70	102
1972-73	446	418	30	102	101
1973-74	500	436	28	101	114
1974-75	444	368	45	114	127
1975-76	478	455	40	127	117
1976-77	490	462	34	117	96
1977-78	546	522	45	96	61
1978-79	501	455	38	61	49

Source : Government of Bangladesh, 1980.

There was serious trouble in the Karnaphuli Complex on 5.2.1972, when 1200 to 1300 previous retrenched workers went to the mill and they went to a confrontation with the job seekers and those who have already appointed in their post (Ahmad, 1978: 102).

A joint statement was issued by Qazi Zafar Ahmad and Haider Akbar Khan (Rono) President and Secretary respectively of Bangladesh Sramik Federation using the Government to take necessary steps immediately to reopen all mills and factories in Tongi area and also to provide relief materials to the workers for their rehabilitation in the second week of February, 1972. On the reopening for Bawa Jute Mills Labour on the 24th February, 1972, Labour Minister of Peoples Republic of Bangladesh Mr. Zahur Ahmad Chowdhury said in his inaugural speech at the reopening of Bawa Jute Mills in February 24, 1972 that the Government would declare the policy of industrial workers very soon. This was followed by a public meeting held at Paltan maidan in the last week of February, 1972 addressed by Qazi Zafar Ahmad. Mr. Ahmad urged the Government to give 80% of the profit of industries to workers, 10% to the management and keep the remaining 10% for itself (Ahmed, 1978: 102).

Being a leftist leader (Pro-Peaking) the above-mentioned demands of Ahmad are quite appropriate when the Government was leading towards socialistic pattern of the economy. But the demand was too much optimistic and early too. Socialism cannot be achieved over night. Not only the opposition but also the Government was stepping hastily without experience and not considering internal as well as global political situation of that time.

In this situation the most elderly mass leader Maulana Abdul Hamid Khan Bhashani issued a statement that many workers who joined their duties after war of liberation were not getting their arrears because banks were not advancing loans to keep the mills running and to purchase raw material and spares. So he urged the Government to take positive steps to resolve the crises. Similarly Mr. Oli Ahad, General Secretary, Bangla Jatiya League appealed to the Government to arrange for arrear salaries to all workers both in public and private sectors as well for the entire period of nine months. Moreover clashes spread out among rival trade unions in the Tongi industrial areas repeatedly. Rowdism, especially in Monnu Textile Mills, Meghna Textile Mills, Nishat Jute Mills became a regular phenomena as a result of inter-union rivalry between Jatio Sramik League (JSL) and Bangladesh Sramik Federation (BSF) Ahmad, 1978: 102-103).

As a result of this unrest all around, coupled with political agitation, Government declared a state of emergency in December, 1974 banning strikes and lockouts, meetings and processions. Under the Emergency, Power Rules, 1975 the functioning of trade unions and the offices of corporations, semi-autonomous organisations, banks, insurance and also Government offices were prohibited, though their registration continued (Hafiz, 1980).

After the proclamation of emergency in January 1975 workers of some textile industries went on massive strike on 21st October, 1975. There was some trouble in Tejgaon industrial area because of the layoff, retrenchment and termination of services. Some amount of unrest was susceptible in Adamjee Jute Mills also. In Dhaka and Comilla areas the managers were manhandled by a group of workers in May 1976. In Meghna Textile Mills there was a 'gherao' and some apprehension of trouble in Ashraf Jute Mills, (Ahmad, 1978: 123-124) [more details in Appendix A and B).

Sheikh Mujibur Rahman, the President of Bangladesh addressed a largely attended labour rally in Tejgaon in March, 1975, where he announced that there would be only one labour front in the country as there could be only one political party named Bangladesh Krishak Sramik Awami League (BKSAL). The BKSAL was announced as the only political party by President Shaikh Mujibur Rahman on 7th June, 1975. According to the constitution of the party, it will have five fronts, namely, Jatio Krishak League, Jatio Mahila League, Jatio Students League, Jatio Juva League and Jatiyo Sramik League and each one of these fronts would be headed by a secretary General and Sheikh Mujibur Rahman would continue to be the Chief of BAKSHAL (Ahmad, 1978: 123).

In the mean time a military coup took place in August 1975. On the 15th August, 1975 Khondakar Mushtaq Ahmed took over after the assassination of Sheikh Mujibur Rahman as the President of the country. President Moshtaq appointed Dr. Khitish Chandra Mondal as the State Minister in Charge of Labour. Later on Zia took over and his Martial Law Government promulgated the Industrial Relations (Regulation) Ordinance, 1975 in December 1975 (sec. 4 of the IRO, 1975). Persons employed as members of the watch and ward or security staff or confidential assistants of any establishment were denied the right to be members or officers of any trade union (sec. 6, ibid). Determining the collective bargaining agent (CBA) through elections was prohibited and a consultative committee was required to be constituted where there was no CBA (sec. 7 and 8, ibid). All industrial disputes were to be settled through peaceful means (sec. 9, ibid). The Government appointed the Industrial Workers Wages and Productivity Commission (IWWPC) to suggest a revised wages scale for the workers of public manufacturing sector in April 1977. The Commission submitted a report in June 1978. And accordingly a new Pay Commission was simultaneously appointed for the other public sector workers and employees (Abbas Ali Khan, 1986 : 96).

The Industrial Relations (Regulations) Ordinance 1975 and the Emergency Power Rules (1975) also affected the structure of trade unions in the country. The Executive order of the 20th July 1977, Industrial Relations Ordinance 1977 relaxed some of the earlier restrictions on trade union activities. The labour policy (1980) has laid the founction of a new structure of trade unions in Bangladesh (Mondal, 1992 : 15). Government's view was that they believe in the process of collective bargaining for resolving industrial disputes and good labour relations. The Government claimed that the policy had been designed to protect the interest of the workers, the management and consumers and the state as well. This policy recognised the right to strike by the workers and that of lockout by employers as instruments of collective bargaining.

All the previous labour policies and the laws introduced subsequently, provided for collective bargaining practices as a means of settlement of disputes, occasional restrictions were imposed on the free play of collective bargaining practices depending on the political situation in the country. The legal provisions for recognitions of collective bargaining agents were made in 1969. The collective bargaining practices did not however take their roots in the pre-independence Bangladesh Labour relations, as the right to strike was granted to the trade unions (Mondal, 1992: 15).

Between 1948 and 1978, 978 strikes and work stoppages were recorded in the Pakistani Bangladesh. Due to these strikes 83 lacs 80 thousand 5 hundred 15 labour days had been lost. In independent Bangladesh 42 lacs 77 thousand 7 hundred 77 labour days had been lost since liberation to 1982 due to industrial unrest in different Government and private productive industrial units (Kaisar, 1992 : 35) see Appendix 1 and II).

The National Federation of Trade Unions (NFTU) consisting of eleven out of sixteen (later joined into Sramik Karmachari Olikya Parishad [SKOP]), had submitted a five point charter of demands that mainly include free trade unionism, stoppage of denationalisation process of public industrial enterprise, wage increase and similar other demands to the Chief Martial Law Administration (CMLA) on Nov. 16, 1982. Later on they also demanded withdrawal of the martial law (Taher, 1997 : 74). SKOP launched various programmes successfully such as rally, torch procession, demonstration, strike, *hartal,* blockade etc. These were observed towards realising their demand at the call of SKOP. The military Government had been compelled to enter into an agreement with SKOP on May 21, 1984 to observe the success of this loose confederation called Sramik Karmochari Oikya Parishad (SKOP). Through this agreement some of the vital trade union rights were revived (Taher, 1977: 74).

The following table 4.3 shows the number of workers in major industries in Bangladesh. From this table it is evident that majority of workers are engaged in jute industry.

Table 4.3 : Number of Workers in Major Industries

Industry	*Number of workers*							
	1982	*1983*	*1984*	*1985*	*1986*	*1987*	*1988*	*1989*
Cotton Mills	80923	84969	190697	199543	226132	231655	7296	72049
Jute Mills	286788	303995	265512	257905	269407	256407	137858	137907
Sugar Mills	24264	25477	25529	27559	27609	27649	21168	21203
Cement, lime and	2125	2221	2647	2692	2746	2746	1483	1483
Glass factories	4543	4770	5293	5391	5731	5821	3001	2041
Match factories	11286	11850	12616	12801	12909	12994	6391	5881
Paper Mills	7605	8027	8883	8973	9071	9168	6037	6087
Engineering	51146	53703	55127	65689	57651	59249	29842	29882
Total	468670	495012	566304	571553	611256	605689	213076	276533

Source : Government of Bangladesh, 1990.

Table 4.4 : Number of Workers Employed in Jute Industries, 1982-1989

Year	*No. of workers engaged In jute industry*	*% to total*	*Total number of workers in major Industries*
1982	286788	61.19	468670
1983	303995	61.41	495012
1984	265512	46.88	566304
1885	257905	45.12	571553
1886	269407	40.07	611256
1987	256407	42.33	605689
1988	137858	46.69	213076
1989	137907	49.86	276533

Source : Compiled from Statistical Year Book, Bangladesh 1990.

Trade Union Movement from 1981 to 1991

From June, 1982, major changes took place in the economic policy of the country, though it originated few years back. President Ziaur Rahman adopted a policy in the late 1970s for a return of substantial part of the nationalized sector to private ownership, while General Ershad, in June, 1982, announced a major programme of decentralisation, notably of the jute and textile industries (Harper, 1987 : 33). The total labour force in 1984-85 was 29.5 million, representing 30.19% of the total population estimated at 97.7 million (Khan, 1990 : 122).

Due to industrial unrest 42 lacs 77 thousand 7 hundred 41 labour days had been lost in 1982. During this period (1981-1991) 5 hundred 21 times production stopped due to direct confrontation between labour and management (Kaisar, 1992 : 35). Industrial disputes took place due to different reasons. The main reason of dispute is wage and other financial issues. The inflation rate in Bangladesh is one of the highest in the world, which demands frequent revision of scale of pay and wages. As a result, 64 percent of the total strike was due to wages. Non-financial matters like leaves, holidays and personal matters like discharge, dismissal, promotion, transfer etc. were said to cause 12 percent of the total number of strikes (Sengupta & Muniruzzaman, 1979 : 15). Although in 1990-91 the nominal wages of both the skilled and unskilled workers in general have increased ten to eleven time of 1969-70 level but real wages have declined by one-fifth to one-fourth Mondal : 19).

The reasons of disputes in a study (Chowdhury, 1983 : 308) have following findings–pressure for the enhanced wages beyond the prescribed norm (31.97%) has been found as number one reason for disputes between the workers and the management. The unfair dismissal and retrenchment of workers (40%) by the management, promotion, job security transfer (21.90%) questionable quality and irregular supply of raw materials (16.18%), misbehaviour of the authority etc. are other important causes.

Industrial disputes take places in Bangladesh for sixteen times in 1983 which causes 3 lacs 92 thousand 6 hundred and 16 man-days lost in 1983 (Kaiser, 1992 : 35). Industrial disputes or conflicts are almost the part of industrial life. But frequent disputes hamper production and distribution process. The number of industrial disputes, workers involved in it and man-days lost since 1947 to 1989 can be seen in notes-1. Industrial disputes some times occur due to political causes (inter and intra-union rivalries, political issues of national interest and sympathetic strikes for causes of specific political party etc.). During 1977-1991 out of 796 cases of industrial disputes 350 (about 44%) cases were causes by so called political factors (Mondal, 1992). Due to industrial disputes, on an average, TK. 105.58 million loss of production occurred during 1980-1991 (Sengupta & Muniruzzaman, 1997 : 17). One hundred and forty two industrial conflicts took place in 1984. Due to this conflict 11 lac 75 thousand one hundred 78 man-days were lost. President H.M. Ershad's regime was full of (Kaiser, 1992 : 35) unrest, which prevailed not only in labour front but also to the political, and student front. It is revealed from different statistics of labour department that indices of trade unions are increasing more rapidly than the number of trade union members. The growth of trade unions is higher than trade unions' membership strength, what are the reasons behind this unnatural growth, i.e., multiplicity of trade unions. Firstly political parties play the most dominant role towards the multiplicity of trade unions in Bangladesh. Different political parties are in competition among themselves to get hold of maximum influence in labour front.

Secondly, The union leaders very often change their party affiliation and try to identify themselves with the party (in the labour front) sponsored by the successive regimes. It seems that there may be two reasons behind this change of identification: (a) Fear of oppressive and repressive measures against the labour leaders may be adopted by the regime in power; and (b) to receive personal gains or favour from the incumbent regime (Bhuyan, 1991 : 153). Not only the labour leaders and even the managers switch over their support and blessing to the unions of ruling party. Managers of public and private enterprises deliberately support ruling party in order to seek benefits from the Government (Hussain *et al.*, 1998 : 146).

The extent of multiplicity may be well understood by the fact that at the national level there are as many as 23 national federations of trade unions, of whom the most are affiliated with different political parties (Taher *et al.*, 1997 : 103). The trade union movement of Bangladesh is highly politicised and the majority of the union is affiliated to political parties. As part of their campaign for the end of Martial Law and for elections, opposition parties and their affiliated unions called a series of general strikes from 1983, often with considerable effect in Dhaka and other large cities and with loss of lives (Harper, 1987 : 34).

From table 4.5, we can have a glance at the political affiliation of trade unions in Bangladesh. Like the following example most of the trade unions of mills and factories have political affiliation or attachment.

There were twenty-three National Federations of Labour in Bangladesh in 1995 with their political affiliation.

Table 4.5 : Distribution of Numbers and Members of National federation of Labour in Bangladesh, 1995

	Name of the Federation	*Number of affiliated unions*	*Membership of affiliated unions*
1.	Bangladesh Jatiya Sramik League	90	50,693
2.	Jatiya Sramik Federation	19	18,300
3.	Bangladesh Trade Union Kendra (CPB)	60	36,881
4.	Bangladesh Sangjukta Sramik Federation L	45	21,272
5.	Bangladesh Trade Union Sangha L	12	19,303
6.	Jatiya Sramik Jote [JSD (Rob)]	12	2,260
7.	Bangladesh Jatiyabadi Sramik Dal (BNP)	218	2,09,748
8.	Bangladesh Sramik Kalyan Federation (JI)	20	9,570
9.	Bangladesh Ganatantrik Sramik Federation (WP)	3	5,937
10.	Banglà Sramik Federation (L)	3	294
11.	Jatiya Sramik League (AL)	144	1,46,972
12.	Bangladesh Trade Union Federation	4	1,648
13.	Bangladesh Free Trade Union Congress (NGO)	19	60,196
14.	Bangladesh Sramik Federation	5	3,639
15.	Jatiya Sramik Party (JP)	142	1,17,844
16.	Jatiya Sramik Forum	11	4,205
17.	Bangladesh Jatiya Sramik Federation (L)	5	2,235
18.	Samajtantrik Jatiya Sramik Front (BSD)	21	2,671
19.	Jatiya Sramik Jote, Bangladesh [JSD (Inu)]	6	1,671
20.	Jatiya Sramik Karmachari Jote	26	5,019
21.	National Trade Union Federation	5	1,799
22.	National Workers' Federation	10	10,467
23.	Bangladesh Workers' Federation	7	1,19,324
	Total	887	8,52,532

Source : Talukder, 1997 : 152-153.

Notes : CPB = Communist Party of Bangladesh; L + = Splinter Leftist Parties, JSD = Jatiya Samajtantrik Dal, BNP = Bangladesh Nationalist Part (Bangladesh Jatiyatabadi Dal), JI = Jamat-I-Islami, WP = Workers' Party, AL = Awami League JP = Jatiya Party, BSD = Bangladesh Samajtrantrik Dal, These political parties are the mentors of the so-called national federation. JSD (Rob) represent a faction of JSD and JSD (Inu) represents another faction of JSD. NGO = Non government voluntary development organization.

So we can come to a conclusion that multiplicity of trade union is a by-product of politicization of unions. There are other factors like attitude of workers and they are hiving less integrity and national feeling. Our workers are ill paid and not educate when their demands and aspirations are not fulfilled by trade unions and their leaders they become frustrated. Then they might have a tendency to switch over from one union to another and from one leader to the other.

The new Martial Law Government headed by H.M. Ershad took some steps to improve the live of industrial workers. The most remarkable of these was almost doubling the wages and salary levels for public sector industrial workers according to the recommendation of the new wage and productivity commission and the new national pay commission appointed in the first half of 1985 (Abbas 1986 : 97). However, the President had been alleged or criticized on the point that this enhancement of salary was motivated. It was one sort of political infiltration among the workers who are generally believed to have a significant influence in the national politics. What ever might be the intention worker were benefited.

With the withdrawal of Martial law at the end 1986, all restrictive provisions on trade unionism and labour relations were also withdrawn correspondingly (Abbas, 1986 : 98). By 1986 there were 2,614 registered trade unions, organised mainly on industrial and occupational lines, and 17 national trade unions. But only 3 percent of the workforce was organised. The 74 percent of unorganized agricultural labourers were little touched by unionism (Harper, 1987 : 34).

One lac seventy five thousand one hundred seventy eight man hour lost in 18 industrial conflicts took place in 1987. Sixteen industrial conflicts was recorded in 1989, while 85 thousand 3 hundred and 95 man hour was lost (Kaisar, 1992: 35).

There were 3625 trade unions in which the total number of members was 14,21,834 in Bangladesh on 31 Dec. 1988. In Dec. 31 1989 number of unions rose to 3908 and the total number of

Table 4.6 : Distribution of Number of Trade Unions and Number of Trade Union Members over the Period 1971-1991

Years	*Number of Unions*	*Number of Members*
1971	1160	45060
1972	2523	682923
1973	3069	859735
1974	3320	946977
1975	3161	921152
1976	3037	895904
1977	3005	873128
1978	3178	1006516
1979	3357	1041080
1980	3613	1097757
1981	3533	1127506
1982	2156	946048
1983	2126	947281
1984	2488	1075496
1985	2593	1090338
1986	2132	1164289
1987	3387	1195704
1988	3625	1421834
1989	3908	1517567
1990	3789	1619008
1991	3956	1642915

Source : Ahmed, M.U. (1997).

members went up to 15,17,567 *(Government of Bangladesh,* 1988 & 1989).

The following table 4.7 shows the growth and development of trade unions and trade union members from 1984 to 1995.

Above table 4.7 shows that although the number of trade unions has been increased to a great extent, the number of trade union members has not increased to the extent of number of trade unions. It is due to multiplicity of trade unions on the one hand, and multiplicity of trade union on the other is due to

Table 4.7 : Distribution of Number of Trade Unions and Number of Trade Union Members – 1984-1991

Years	*Number of Registered Trade Unions*	*Indices of Trade Unions (Base=1984)*	*Number of Members of Trade Unions*	*Indices of members of Trade Unions (Base=1984)*
1984 (31st Dec.)	2488	100	1075496	100
1985 (31st Dec.)	2596	104.22	1090338	101.38
1986 (31st Dec.)	3132	125.88	1164279	108.26
1987 (31st Dec.)	3387	136.13	1195704	111.18
1988 (31st Dec.)	3625	145.70	1421834	132.20
1989 (31st Dec.)	3708	149.04	1517567	141.10
1990 (31st Dec.)	3789	152.29	1619008	150.54
1991 (31st Dec.)	3956	159.00	1642915	152.76

Source : M.A. Taher et al., 1998 : 108.

political affiliation. There are some other causes too, inter-regional groupings and personality clashes among the labour leaders. For example, in Bangladesh Bank, the Central Bank of the country, four trade unions are in operation in Adamjee Jute Mills Ltd. The Government of Bangladesh has been compelled to close down the mill, due to recurring huge losses. This jute mill was the biggest in Asia and it had a worldwide reputation in Jute sector of the world. But there cannot be more than three unions in an enterprise as per labour law of the country. It is a matter of regret that most of the organizations are violating the provisions of labour laws of the country.

The Movement During 1991-2000

Though in a mass movement, President Hussain Muhammad Ershad was ousted and the elected Government came in power in 1991. So democratic environment was created within the country. The right to form or join any association or organisation had been ensured as democratic rights restored. So union activity started in full swing.

There were 2488 registered trade unions with 10,75,469 members in Bangladesh in 1984, whereas it stands 4514 trade unions with 17,02,112 members in 1995 (Taher *et al.*, 1998 : 102).

Table 4.8 : Distribution of Number of Trade Unions and Number of Trade Union Members over the period 1992-1995

Years	*Number of Unions*	*Number of Members*
1992	4056	1648783
1993	4238	1674945
1994	4484	1681694
1995	4796	1720679

Source : Ahmed, M.U. (1997).

Table 4.8 represents that the number of trade unions is increasing rapidly, but the number of members is not increasingly rising accordingly. This tendency leads to the multiplicity of trade unions.

Upto 31st December 1998 there were 5694 trade unions in operation with 18,23,561 members. Additional 337 unions had been registered with 50,144 members in 1999. But 116 unions had been cancelled, which had 17,011 members. So upto 31st December 1999 total number of trade unions was 5915 with 18,56,694 members.

Table 4.9 : Classification of Trade Unions According to Number of Members on 1998

According to Number	*Number of Unions*	*Number of Members*
01-50	145	4063
51-100	107	6054
101-200	63	11059
201-300	30	6087
301-400	06	1864
401-500	05	2285
501-600	02	1050
601-1000	07	5263
1001-2000	02	2342
2001-5000	01	2002
5000-Above	01	12172
Total	369	54241

Source : Government of Bangladesh, 1998 : 36.

We see the same scenario in the above table 145 unions having members 01-50. Only two unions having membership above one thousand and only one union has more than twelve thousand members. So there is a trend of forming small unions, which indicates sign of disunity and multiplicity of unions.

Adamjee Jute Mill was renowned in the trade union history of this country. There were about 28 thousand workers and four thousand employess employed in the mill in 1992. In the pre-liberation period workers of this mill played a pivotal role in the different anti-Government movements. But after that a group of corrupt leaders used the workers in terrorist and other activities derogatory to the workers interest. Corrupt leaders used to take percentage in jute purchase, sale, submission of tender and appointment of new workers. In this way they made money and destroyed the workers' interest and finally jeopardized the total objective of the trade union movement. Management was also corrupt and inefficient. About 110 workers were assassinated almost all in inter-union or intra-union rivalry from 1972 to 1992 (Kaiser, 1992 : 92). Not only in Adamjee Jute mills, but also in all the mills including banks, transport, etc. were the same. One of the biggest sectors is road transport unions, where about four hundred big and small unions were in operation. Three federations were united and formed Bangladesh Road Transport Workers' Federation in 1990. But all over the transport sectors, unions were also corrupt. They used to extort money illegally from all the bus and truck terminus and which were distributed upto top level. In this way all the leaders used the unions as their media of corruption (Kaiser, 1992).

The following table shows the distribution of trade unions and number of trade union members. Again we can see the number of trade unions is increasing but the number of members is not increasing proportionately.

Number of workers of major industries in public and private sector in Bangladesh is shown in table 4.11.

Table 4.10 : Distribution of Number of Trade Unions and Number of Trade Union Members–1992-1995

Years	*Number of Registered Trade Unions*	*Indices of Trade Unions (Base=1984)*	*Number of Members of Trade Unions*	*Indices of members of Trade Unions (Base=1984)*
1992 (31st Dec.)	4065	163.38	1648783	153.30
1993 (31st Dec.)	4238	170.34	1674949	155.74
1994 (31st Dec.)	4484	180.23	1681690	156.36
1995 (31st Dec.)	4514	181.43	1702112	158.26

Source : Taher et at., 1998 : 108.

Garment industry has been developed in Bangladesh in tremendous way within a very short span of time. This is a labour-intensive industry and the advantage is that labour is very cheap in Bangladesh in comparison to western and even to many Asian countries. Before liberation there were only one garment factory in Bangladesh. There were only 8 factories in 1997 whereas the number rose to 587 in 1984. The number of garment factories stood at 2650 in 1998 and at present the number is 3,300 (BNGWF, 1999-2003 : 2003).

Garments is the highest source of foreign earnings of Bangladesh as jute was the leading source of foreign exchange of the then Pakistan period. A clear picture of export earning of principal commodities of Bangladesh is shown in table 4.12.

Trade union movement in garments industry is very week. It is weaker than other sectors even. There are 8 country-wide registered trade union Federations. There are 9 federations registered as division based. Another 5 registered federation are combined with jute, textile and leather sector. Apart from these, there are 6 unregistered federations in this sector. There are three alliances in Garment sector. These are : (a) Bangladesh Garment Workers Unity Council; (b) Bangladesh Garments Workers and Employees Unity Council; and (c) Bangladesh National Coordinating Committee (BNCC) affiliated with (ITGLWF. The National Garment Workers Federation (NGWF)

Table 4.11 : Number of Workers in Major Industries

Industries	1996-97	1997-98	1998-99	1999-2000
Jute				
Production Worker	112554	112673	112508	111075
Other employees	14855	14690	14856	14385
Total	127409	127363	127364	125460
Cotton				
Production Worker	41508	41582	40969	414287
Other employees	7318	7395	7540	7585
Total	48826	48977	48509	49013
Paper				
Production Worker	3601	3484	3392	3310
Other employees	2860	2767	2709	2648
Total	6461	6251	6101	4994
Steel				
Production Worker	1668	1600	1509	1550
Other employees	554	538	506	492
Total	2222	2138	2015	2042
Cement				
Production Worker	816	884	1070	983
Other employees	571	672	780	602
Total	1387	1556	1850	1585
Fertiliser				
Production Worker	3080	3134	3128	3109
Other employees	4128	4066	4050	4091
Total	7208	7200	7178	7200
Petroleum				
Production Worker	436	429	406	379
Other employees	296	300	321	356
Total	732	729	727	735
Print and Varnish				
Production Worker	234	250	260	273
Other employees	279	286	293	319
Total	513	536	553	592
Ready Made Garment (For export mainly)				
Employees* (in million)	1.40	1.50	1.51	1.50 (e)
Male	0.14	0.15	0.15	0.15 (e)
Female	1.26	1.35	1.36	1.35 (e)
Number of Garment factories	2503	2726	2963	2963

* The total number of technical and production workers and other employees. e : estimated

Source : Government of Bangladesh, 2000 : 164.

Table 4.12 : Export of Principal Commodities of Bangladesh (1995-96 to 1998-99)

(Million Taka)

Commodities	*1995-96*	*1996-97*	*1997-98*	*1998-99*
Total Exports	**1144521**	**171554**	**229408**	**245620**
1. Prawns and shrimps	11803	10781	11745	12117
2. Tea	1254	1380	2150	1885
3. Spices	2	6	1	0.4
4. Raw hides and skins	2	3	1	1
5. Raw Jute	2968	4344	3776	3527
6. Jute Yarn	2405	2854	2684	2970
7. Jute mfg. Total	10360	10531	9528	19719
(a) Hessian	1956	2108	1812	2309
(b) Sacking	35	45	14	18
(c) Carpet backing cloth	1459	1156	1304	1089
(d) Others	6910	7222	6398	7303
8. Leather and leather manufacture	6108	5366	4809	6793
9. Ready made garments	90595	114793	171660	190146
10. Handicraft	135	66	150	200
11. Others	18889	21430	22904	17262

Source : Government of Bangladesh, 2000 : 256.

is the most important trade union federation of garment workers in Bangladesh. The federation was established in 1984. There are 28 registered trade unions (plant unions) affiliated with NGWF. Beside these, the federation had 1016 factory committee. From 1984 the federation is involved in all the important movements including strikes in the Garment sector (Activity Report, 1999-2003).

Table 4.13 : Number of Employees in Ready Made Garments

(in millions)

Sex	*1996-97*	*1997-98*	*1998-99*	*1999-2000*
Male	0.14	0.15	0.15	0.15
Female	1.26	1.35	1.36	1.36
Total	1.40	1.50	1.51	1.51

Source : Bangladesh Statistical Pocket Book 2000.

In garments sector ratio of female workers is much higher than the male. Women workers are the backbone of garments industry, but their economic and social status is at stake. Even the management does no ensure the security of women workers. Women workers faced rapes and sexual harassment outside the factories and sometimes inside the factories. Women workers are not given the equal wage, equal rights and promotional facilities (ibid.).

By and large, this is the total scenario of trade union movement of Bangladesh. At present, there are 26 federations in existence in Bangladesh and it is the result of multiplicity of trade unions. The table (Bangladesh Institute of Labour Studies [BLIS]) will show the names of National Federations, their office-bearers, number of unions attached with the respective federation, number of unions affiliated with and number of members with each union (see appendix-III).

Summary

In this chapter we have attempted to discuss the growth and development of trade unions and their movement since liberation in 1971 to 2000 with the frequent political changes that took place within the period. Trade union movement got a new fillip in a newly liberated country and under a new economic order. But it is marked that they just follow the predecessors. The very practice of multiplicity of trade unions, inter-union rivalry, leadership conflict, politicisation of unions, low academic background of trade union leaders and corrupt leadership of trade unions have become more acute than before.

For a couple of years the trade unions have lost their glorious past. The union movements are now regarded as a disturbing element in the development process. Donor countries now-a-days have been imposing constant pressure on us to impose restriction on those union activities that hamper growth and productivity of the manufacturing enterprises.

REFERENCES

Ahmad, M.U., (1997), *Productivity-linked Industrial Wages Agreement in Bangladesh : An Overview,* Report Prepared for ILO (SATT), New Delhi.

Akkas, Md. Ali, (1998), Labour Management Cooperation : The Bangladesh Perspective, *Dhaka University Journal of Business Studies,* Vol. 19 (2).

Akkas, Md. Ali and Hussain, Ferdous, (1998), Characteristics and Weaknesses of Trade Unionism : Bangladesh Perspective, *Social Science Review,* Vol. XV, No. 2, Department of Sociology, University of Dhaka.

Alam, Md. Fashiul (1983), Industrial Disputes Laws in Bangladesh : A Study in their Development Perspective, *The Cost and Management,* September-October.

Bangladesh National Garments Workers Federation (BNGWF) (2003): *Activity Report,* 1999-2003.

Bhuyan, M. Sayfullah, (1986), Trade Union Movement, in Chakravarty, S.R. and Narain, Virendar (eds.) : *Bangladesh,* Vol. II. Domestic Politics, South Asia Studies, Series 13, South Asian Publishers, New Delhi.

Bhuyan, M. Sayfullah, (1991), The Trade Union Politics in Bangladesh: A Study of Textile Mills Workers Union, *Dhaka University Studies,* Part C, Vol. 12 (1).

Chowdhury, A. Mannan, (1983), *The Settlement of Disputes – A Study of the Jute Industry of Bangladesh.*

Government of Bangladesh, (1980), *Statistical Year Book.*

Government of Bangladesh, *Bangladesh Economic Survey,* 1984-85.

Government of Bangladesh, *Bangladesh Labour Journal,* 1988 and 1989.

Government of Bangladesh, (1990), *Statistical Year Book.*

Government of Bangladesh, *Bangladesh Labour Journal,* 1999, Part 19.

Harper, John F. (ed.) (1987) : *Trade Unions of the World,* Gale Research Centre, Detroit.

Ibid. Section 6.

Ibid. Section 7 and 8.

Ibid. Section 9.

Industrial Relations Ordinance, 1969, with commentary M. Shafi, Vol. 1969, Bureau of labour Publications, Karachi.

Industrial Relations ordinance, 1975, Section 4.

Kaisar, Ashraf, (1992), Business in the Name of Trade Union (in Bengali), Bichitra, 7th August.

Quoted in Khan, M.M. and Ahmad, M., (1980), *Participate Management in Industries.*

Sengupta, Palash and Muniruzzaman, A.K.M. (1997): Industrial Disputes in Bangladesh : Trends, Causes and Impact, *South Asian Journal of Management,* April-June.

Taher, Md. Abu, (1997), Conflict and Accommodation : SKOP, Govt. and the Private Employees in Bangladesh.

Taher Md. Abu, Muniruzzaman and Mahmood, M.M.H., (1998), Multiplicity of Trade Unions in Bangladesh : An Exploratory Study, *Islamic University Studies,* Part C, Vol. 1, No. 1, June, Kushtia, Bangladesh.

Talukder, M.A.S., (1997), *Labour and Industry* (in Bengali) Nasa Publishing Co., Dhaka.

The Daily Star, Dhaka, January 8, 1998.

APPENDIX - 1

Trends of Intensity of Industrial Disputes (1972-1994)

Year	*No. of dispute (D)*	*No. of Workers involved (W)*	*No. of Man-days lost (L)*	*Index No. of disputes Base 1972=100*	*Index of Workers Involved Base 1972=100*	*Index of Man-days lost Base 1972=100*
1972	39	43615	126000	100.00	100.00	100.00
1973	58	35027	285177	148.72	80.31	226.23
1974	32	57387	231736	82.05	131.58	183.92
1975	02	28327	162000	05.13	64.95	128.57
1976	05	14517	256618	12.82	33.28	20.33
1977	22	76675	81715	56.41	175.80	64.85
1978	89	113209	662332	228.21	258.56	525.66
1979	96	114248	647629	246.15	261.95	513.99
1980	104	164032	1160436	266.67	376.09	920.48
1981	80	117031	1198460	205.13	268.33	268.33
1982	55	21788	238658	141.03	49.96	189.41
1983	16	175787	392616	41.03	403.02	311.60
1984	142	481004	1144817	364.10	1102.84	908.58
1985	95	198118	284920	243.59	454.24	226.13
1986	46	105977	2079671	117.95	242.98	1650.53
1987	18	88795	175278	46.15	203.59	140.11
1988	09	28874	49398	23.08	66.20	39.20
1989	16	60513	85395	41.03	138.74	67.77
1990	05	14554	28479	12.82	33.37	22.60
1991	03	454	29811	07.69	01.04	23.66
1992	11	6190	29741	28.21	14.19	23.60
1993	11	6101	16029	28.21	13.99	12.72
1994	02	5641	2421	05.13	12.93	01.92
Period Average 1972-1994	41.57	85124.52	397319	106.58	195.17	315.33

Source : Sengupta and Moniruzzaman, 1997 : 14.

APPENDIX - II

Trends of Industrial Disputes for Disposal (1972-1994)

Year	*Total No. of disputes not involving stoppage of work*	*Total No. of disputes involving stoppage of work*	*Total No. of disputes referred to labour courts*	*Total No. of disputes referred to labour Tribunal*
1972	720	39	3377	42
1973	1645	58	4056	430
1974	883	32	4910	237
1975	469	02	4532	23
1976	311	05	4048	19
1977	276	22	4032	52
1978	408	89	4010	51
1979	361	96	3290	41
1980	570	104	3876	38
1981	673	80	4893	58
1982	408	55	6704	42
1983	420	16	6923	34
1984	837	142	5605	175
1985	872	95	4654	156
1986	522	46	5256	25
1987	343	18	5732	30
1988	573	09	5538	38
1989	2822	16	533	22
1990	564	05	5499	24
1991	1293	03	5264	31
1992	1204	11	4374	52
1993	272	11	4622	38
1994	218	02	4482	63

Source : Sengupta and Moniruzzaman, 1997 : 11.

APPENDIX - III

List/Statements of National Federations

Sl. No.	*Name of the Federation and Address*		*Name of the President & Secretary*	*Number of affiliated unions*	*Number of members of affiliated union*
1	2		3	4	5
1.	Jatio Sramik Federation	(i)	Mr. Shafiqur Rahman Majumdar	19	18,280
	76/1, Q, Uttar Jatrabari, Bibir Bazar, Dhaka	(ii)	Mr. Jalal Ahmad		
2.	Bangladesh Trade Union Kendro	(i)	Mr. Nurul Islam	60	50,180
	23/2, Topkhana Road, Dhaka	(ii)	Dr. Wajedul Islam Khan		
3.	Bangladesh Sanjukta Sramik Federation	(i)	Mr. Mukhlesur Rahman	119	21,272
	23/2, Topkhana Road, Dhaka	(ii)	Mr. Sharitullah		
4.	Bangladesh Trade Union Sangha	(i)	Mr. Khalilur Rahman	12	19,933
	31/32, P.K. Roy Road, Bangla Bazar, Dhaka	(ii)	Mr. Chowdhury Ashiqul Alam		
5.	Jatio Sramik Jote	(i)	Mr. Meshbahuddin Ahmad	12	2,260
	20/4, Jai Kali Mandir, Dhaka	(ii)	Mr. Abdul Wahed		
6.	Bangladesh Jatiotabadi Sramik Dal	(i)	Mr. Alhoj Abdullah Al Noman	258	2,32,899
	28/1, Naya Paltan, Dhaka	(ii)	Mr. Nazrul Islam Khan		
7.	Bangladesh Mukta Sramik Federation	(i)	Mr. Md. Noshu Mian	128	75,693
	68/2, Purana Paltan Dhaka	(ii)	Mr. Mujibur Rahman		
8.	Bangladesh Sramik Kalyan Federation	(i)	Mr. Mujibur Rahman	33	34,522
	435, Elephant Road (Baro Mogbazar), Dhaka	(ii)	Prof. Harunur Rashid Khan		

Contd....

APPENDIX - III (Contd....)

Sl. No.	Name of the Federation and Address	Name of the President & Secretary	Number of affiliated unions	Number of members of affiliated union
1	2	3	4	5
9.	Jatio Sramik Federation, Bangladesh 31/F, Topkhana Road, Dhaka	(i) Mr. Abul Bashar (ii) Mr. Shah Alam	11	15,881
10.	Jatio Sramik League 23, Bangabandhu Avenue, Dhaka	(i) Mr. Abdus Salman Khan (ii) Mr. Ahsan-ullah Master	144	1,46,972
11.	Bangladesh Trade Union Federation 12/1, Quazi Nazrul Islam Sarani, Mymenshing Road, Bangla Motor, Dhaka	(i) Mr. Shah Atiul Islam (ii) Mr. Sadekur Rahman	04	1,648
12.	Bangladesh Free Trade Union Congress 6/A, 1/19, Mirpur, Dhaka	(i) Mr. S.M. Alam Mendu Mian (ii) Mr. Mamunur Rashid Chowdhury	33	1,58,693
13.	Bangladesh Sramik Federation 14, Shaheed Captain Mansur Ali Saroni, Dhaka	(i) Mr. Gholam Sattar (ii) M.A. Mannan	05	3,639
14.	Jatio Sramik Party 60, Dilkusha Commercial Area, Dhaka	(i) Mr. Shah Md. Abu Jafar (ii) Advocate Delwar Hussain	142	1,17,844
15.	Bangladesh Jatio Sramik Forum 12/1, Kabi Jasimuddin Road, Uttar Kamalapur, Matijhil, Dhaka	(i) Mr. M. Jahangir Alam (ii) Mr. Abdul Mannan Siddiky	14	17,885
16.	Bangladesh Jatio Sramik Federation 15/A, Purana Paltan, Dhaka	(i) Mr. Tofazzal Hussain (ii) Mr. Rafiquzzaman	05	2,235

Contd....

APPENDIX - III (Contd....)

Sl. No.	Name of the Federation and Address		Name of the President & Secretary	Number of affiliated unions	Number of members of affiliated union
1	2		3	4	5
17.	Samajtantrik Sramik Front	(i)	Mr. Abdullah Sarkar	16	2,285
	23/2, Topkhana Road, Dhaka	(ii)	Mr. Mahbubur Rahman Ismail]		
18.	Jatio Sramik Jote, Bangladesh	(i)	Mr. Abdul Kader Howlader	20	7,130
	35/36, Bangabandhu, Avenue Dhaka	(ii)	Mr. Badal Khan		
19.	Jatio Sramik Karmochari Jote	(i)	Mr. Rafiqul Islam	26	5,019
	211, Madhya Basabo Railgate				
	(East of Railgate), Dhaka	(ii)	Mrs. Fatema Begum		
20.	National Trade Union Federation	(i)	Mr. Mushaf Hussain	05	9,791
	B.T.C. Union Office, Mahakali, Dhaka	(ii)	Mr. Abu Bakar Mian		
21.	National Workers Federation	(i)	Mr. Rahul Amin Majumdar	10	10,467
	48, Purana Paltan Dhaka	(ii)	Mr. Anisul Haque		
22.	Bangladesh Workers Federation	(i)	Not Available	07	1,19,324
	40, Siddheswari (Ayesha Manzil), Dhaka	(ii)	Not Available		
23.	Bangladesh Sramik Sanghoti Federation	(i)	Mr. Md. Alam Khan	12	3,664
	20/25, North South Road, Habib Market, Dhaka	(ii)	Mr. Rahul Amin		
24.	Bangladesh Jatio Sramik Karmochari Parishad	(i)	Mr. Alhaj M.A. Azia Shisti	05	8,465
	13/1/C, 2 K.M. Das Lane, Golapbagh, Dhaka	(ii)	Mr. Hedayatul Islam		

Contd....

APPENDIX - III (Contd....)

Sl. No.	Name of the Federation and Address	Name of the President & Secretary	Number of affiliated unions	Number of members of affiliated union
1	2	3	4	5
25.	Jatio Samajtantrik Sramik Jute 35/36, Bangbandhu Avenue, Dhaka	(i) Mr. Mumtazuddin (ii) Mr. Abuk Khan Sajjad	11	3,840
26.	Bangladesh Jatio Sramik Jute 35/36, Bangabandhu Avenue, Dhaka	(i) Mrs. Shirina Akhtar (ii) Mr. Nur Khan	05	2,421

Source : Compiled from the Statistics provided by *Bangladesh Institute of Labour Studies*, 2003.

Chapter - 5

PERFORMANCE OF TRADE UNION IN PUBLIC SECTOR JUTE MILLS

Introduction

In Bangladesh, nationalised jute mills comprise the largest mills in terms of production, export and employment generation. This chapter gives an exposition of performance of trade union in public sector jute mills of Bangladesh with the emphasis of detailed analysis of two jute mills, namely, Jessore Jute Industries (JJI) and Rajshahi Jute Mills (RJM). We have collected secondary data to represent macro-economic scenario of production export, internal consumption of jute goods of Bangladesh. Manpower position of different types of jute industry (especially the nationalised and private jute mills) has also been portrayed. Opinion survey has been conducted and according to our purpose we have conducted survey with the workers (80 workers each and in total 160 workers), trade union leaders and mill's executives and presented them in tabular forms. We have also tested homogeneity and/or indifference of two mills on the basis of age, education, parental occupation, role of trade unions, management's attitude towards workers, ways of presenting demands, negotiation process, volume of disputes, settlement of disputes, cause of disputes, nature of workers cooperation, etc. with the help of χ^2 (Chi-square).

Short History of the Two Jute Mills

There was no Jute mill in the then East Pakistan (now Bangladesh) up to 1947. In 1950s, jute mills took its start in Narayanganj, Dhaka and Daulatpur, Khulna under private initiatives. Around 1960, the then East Pakistan Industrial Development Council (EPIDC) established twelve jute mills in Bangladesh having 8000 looms. During that time, there was no jute mill in northern districts. Virtually decisions were taken to establish two jute mills in northern areas one at Rajshahi and another at Sirajganj. In 1969, Rajshahi Jute Mill (RJM) was established with a view to (i) use the locally available raw jute, (ii) to solve unemployment problem, (iii) to earn huge foreign exchange, and (iv) facilitate the rapid industrialisation programme. The site was selected due to low cost of land, convenience in transport of raw materials, cheap labour supply etc. Rajshahi jute mill was established at Shampur, Rajshahi, 8 km. away from Rajshahi town.

Jessore Jute Industries (JJI) was established in 1970 in Rajghat, with similar facilities in raw jute, labour supply available, land and Table 5.1 shows that Bangladesh is the leading exporting country of jute goods in the world and India occupies the second position.

Table 5.1 : World Statistics of Export (000 MT)

Country	*1995-96*	*1996-97*	*1997-98*	*1998-99*	*Avg.*	*SD*	*CV*	*AGR*
World Total	820.6	728.7	719.2	716.4	746.23	43.18	5.79	114.54
Bangladesh	427.9	409.6	343.9	432.2	403.40	35.38	8.77	99.01
India	229.4	180.7	245.4	173	207.13	30.92	14.93	132.60
China	29.9	16.7	14.7	6.9	17.05	8.27	48.52	433.33
Thailand	24.7	11.7	9.8	6.7	13.23	6.86	51.88	368.66
Pakistan	0	0	0	0	0.00	0.00	-	-
Nepal	10	10	10	10	10.00	0.00	0.00	100.00
Rest of the world	98.7	100	95.4	87.6	95.43	4.82	5.05	112.67

Source : Annual Jute Goods Statistics 2000, Bangladesh.

Table 5.1 Shows that Bangladesh is the leading exporting country of jute goods on the World and India occupres the second position.

Table 5.2 : Production of Jute Goods in Bangladesh (000 MT)

Organisation	*1996-97*	*1997-98*	*1998-99*	*1999-2000*	*Avg.*	*SD*	*CV*	*AGR*
BJMC	271.53	255.29	236.19	234.9	249.478	17.4084	6.977951	115.594
BJMA	139.22	153.59	132.06	104.2	132.268	20.743	15.6826	133.608
BJSA	119	132.19	147.77	149.6	137.14	14.3969	10.49794	79.5455
Bangladesh Total	529.75	541.07	516.02	488.7	518.885	22.58	4.351632	108.4

Source : Annual Jute Goods Statistics 2000, Bangladesh.

Table 5.2 Portrays that the average Production of BJMC and BJMA were 249.478 and 132.268, MTs SD was 17.4084 & 20.743, CV was 6.977951 & 15.6826 and AGR 115.594 & 135.608 during 1996-2000 respectively.

Table 5.3 : Total Export (000 MT) of Bangladesh Jute During 1996-2000

Organisation	*1996-97*	*1997-98*	*1998-99*	*1999-2000*	*Avg.*	*SD*	*CV*	*AGR*
BJMC	192.85	160.27	223.84	201.6	194.64	147.62	75.84	134.69
BJMA	91.06	76.2	78.97	67.5	78.43	162.70	207.44	206.25
BJSA	109	116.49	130.57	134.4	122.62	163.60	133.43	88.54
Bangladesh Total	392.91	352.96	433.38	403.5	395.69	173.77	43.92	131.29

Source : Annual Jute Goods Statistics 2000, Bangladesh.

Table 5.3 focuses that the average exports of BJMC and BJMA were 2194.64 & 78.43, MT, SD was 147.62 & 162.70, CV was 75.84 & 207.44 and AGR 134.69 & 206.25 during 1996-2000 respectively.

Table 5.4 : Looms Installed and Operated (BJMC & BJMA)

Period	*Installed (Operated)*					*Operated*				
	Hessian	*Sacking*	*CBC*	*Others*	*Total*	*Hessian*	*Sacking*	*CBC*	*Others*	*Total*
1989-90	14955	8420	2498	147	26020	12589	7351	1773	98	21811
	(57.47%)	(32.36%)	(9.60%)	(0.56%)	(100%)	(57.72%)	(33%)	(8.13%)	(0.45%)	(100%)
1990-91	14955	8420	2498	147	26020	12111	5799	1943	91	19944
	(57.47%)	(32.36%)	(9.60%)	(0.5%)	(100%)	(60.73%)	(29.08%)	(9.74%)	(0.45%)	(100%)
1991-92	15029	8346	2498	160	26033	11051	6390	1991	85	19517
	(57.73%)	(32.06%)	(9.60%)	(0.61%)	(100%)	(56.62%)	(32.74%)	(10.20%)	(0.44%)	(100%)
1992-93	15004	8371	2498	169	26042	9694	7192	1915	97	18898
	(57.61%)	(32.14%)	(9.60%)	(0.65%)	(100%)	(51.30%)	(38.06%)	(10.13%)	(0.51%)	(100%)
1993-94	14985	8390	2498	174	26047	9512	6312	1758	108	17690
	(57.33%)	(32.21%)	(9.59%)	(0.67%)	(100%)	(53.77%)	(35.68%)	(9.94%)	(0.61%)	(100%)
1994-95	14141	8095	2363	152	24751	9669	6180	1755	98	17702
	(57.13%)	(32.71%)	(9.55%)	(0.61%)	(100%)	(54.62%)	(34.91%)	(9.92%)	(0.55%)	(100%)
1995-96	14357	8124	2355	155	24991	8502	6246	1739	89	16576
	(57.45%)	(32.51%)	(9.42%)	(0.62%)	(100%)	(51.29%)	(37.68%)	(10.49%)	(0.51%)	(100%)
1996-97	14946	8424	2420	155	25945	8004	6666	1546	83	16299
	(57.61%)	(32.47%)	(9.32%)	(0.60%)	(100%)	(49.11%)	(40.90%)	(9.49%)	(0.50%)	(100%)
1997-98	14946	8424	2420	146	25936	7835	6930	1503	85	16353
	(57.63%)	(32.48%)	(9.33%)	(0.56%)	(100%)	(47.91%)	(42.38%)	(9.19%)	(0.52%)	(100%)

Contd....

Table 5.4 : (Contd....)

Period	*Installed (Operated)*					*Operated*				
	Hessian	*Sacking*	*CBC*	*Others*	*Total*	*Hessian*	*Sacking*	*CBC*	*Others*	*Total*
1998-99	14905	8485	2420	154	25964	7830	6314	1334	73	15551
	(57.41%)	(32.68%)	(9.32%)	(0.59%)	(100%)	(50.35%)	(40.60%)	(8.58%)	(0.47%)	(100%)
1999-2000	14329	8083	2282	154	24848	7845	5683	1340	77	14945
	(57.67%)	(32.53%)	(9.18%)	(0.62%)	(100%)	(52.49%)	(38.03%)	(8.97%)	(0.51%)	(100%)
BJMC	8592	4508	1421	95	14616	6245	3828	1089	48	11210
	(58.78%)	(30.84%)	(9.72%)	(0.66%)	(100%)	(55.71%)	(34.15%)	(9.71%)	(0.43%)	(100%)
BJMA	5737	3575	861	59	10232	1600	1855	251	29	3735
	(56.07%)	(34.94%)	(8.41%)	(0.58%)	(100%)	(42.84%)	(49.67%)	(6.72%)	(0.77%)	(100%)
Average	13606.23	7666.54	2233.23	143.62	23649.62	8652.85	5903.54	1533.62	81.62	16171.62
SD	2817.98	1561.99	482.94	30.30	4885.84	2689.59	1438.86	450.84	20.81	4391.38
CV	20.71	20.37	21.62	21.10	20.66	31.08	24.37	29.40	25.50	27.15
AGR	260.68	235.52	290.13	249.15	254.30	786.81	396.28	706.37	337.93	583.96
Annual Growth Rate	-39.3	-12.28	-17.5	-0.24	-69.32	-498.93	-52.53	-62.02	-2.04	-615.51

Source : Annual Jute Goods Statistics 2000, Bangladesh.

Average installed loom were 23649 during 1989-90 to 2000.

Average operating loom were 16171.62 during and 89-90 to 2000.

Figures within the parentheses indicate percentage to toal.

Table 5.5 : Weaving Production & Raw Jute Consumption (BJMC & BJMA)

Period	Hessian		Sacking		CBC		Carpet	Others	Total Prod.	Raw Jute
	000' Meter	Ton	000' Meter	Ton	000' Meter	Ton	Ton	Ton		Consumption Million Bales of 180 Kg.
1989-90	637458	177687 (33.58)	622627	268213 (51.44)	91613	67449 (12.75)	711 (0.13)	15045 (2.85)	529106 (100.00)	2.96
1990-91	568385	158479 (36.51)	456789	202635 (46.70)	80518	57179 (13.18)	543 (0.13)	15085 (3.48)	433921 (100.00)	2.468
1991-92	475348	131858 (31.67)	452121	200764 (48.22)	83510	61001 (14.65)	818 (0.20)	21936 (5.26)	416377 (100.00)	2.344
1992-93	418458	116295 (26.09)	548450	244756 (54.91)	78126	56344 (12.64)	1356 (0.30)	27027 (6.06)	445778 (100.00)	2.504
1993-94	410456	115069 (27.28)	509309	223719 (53.04)	70642	51710 (12.26)	3322 (0.79)	27986 (6.63)	421806 (100.00)	2.407
1994-95	0	117550 (27.64)	0	214487 (50.42)	0	50169 (11.79)	2640 (0.62)	40519 (9.53)	425365 (100.00)	2.401
1995-96	0	100721 (24.87)	0	209551 (51.74)	0	48470 (11.97)	3083 (0.76)	43156 (10.66)	404981 (100.00)	2.32
1996-97	0	93055 (22.65)	0	225456 (54.89)	0	44444 (10.82)	2780 (0.68)	45015 (10.96)	410750 (100.00)	2.3
1997-98	0	88108 (21.55)	0	229994 (56.25)	0	40293 (9.85)	2104 (0.51)	48383 (11.84)	408882 (100.00)	2.282

Contd...

Table 5.5 : (Contd....)

Period	*Hessian*		*Sacking*		*CBC*		*Carpet*	*Others*	*Total Prod.*	*Raw Jute*
	000' Meter	*Ton*	*000' Meter*	*Ton*	*000' Meter*	*Ton*	*Ton*	*Ton*		*Consumption Million Bales of 180 Kg.*
1998-99	0	88907	0	198804	0	37666	1347	41531	368255	2.118
		(24.14)		(53.98)		(10.23)	(0.37)	(11.28)	(100.00)	
1999-2000	0	90446	0	180449	0	36883	1349	30017	339144	1.941
		(26.67)		(53.20)		(10.88)	(0.40)	(8.85)	(100.00)	
BJMC	0	71664	0	126850	0	28002	376	8052	234944	1.357
		(30.50)		(53.99)		(11.92)	(0.16)	(3.43)	(100.00)	
BJMA	0	18782	0	53599	0	8881	973	21965	104200	0.584
		(18.02)		(51.44)		(8.52)	(0.94)	(21.08)	(100.00)	
Average	193085.00	105278.54	199176.62	198405.92	31108.38	45268.54	1646.31	29670.54	380269.92	2.15
SD	250274.85	37939.40	254981.12	52776.59	39578.03	14820.73	979.08	12554.42	102248.83	0.57
CV	129.62	36.04	128.02	26.60	127.23	32.74	5947	42.31	26.89	26.54
AGR	155.30	946.05	122.25	500.41	129.69	759.48	73.07	68.50	507.78	506.85
Annual Growth Rate		-8241.55		-3811.11		-2091.51	117.03	2828.46	-12016.7	-0.065

Source : Annual Jute Goods Statistics 2000, Bangladesh.
Annual total production was 380269.92 during 1990-2000.
Annual total Raw Jute consumption 2.15 (Million Bales of 180 kg.) during 1990-2000.
Figures within the parentheses indicate percentage to total.

Table 5.6 : Weaving Production in India

Period	*Hessian*		*Sacking*		*CBC*	*Others*	*Total*
	000' Meter	*Ton*	*000' Meter*	*Ton*	*000' Meter*	*Ton*	*Ton*
1989-90	1088045	340.8	1552326	716.3	30.0	258.8	1345.9
		(25.32)		(53.23)	(2.23)	(19.23)	(100.00)
1990-91	1086445	340.3	1712478	790.2	19.9	271.4	1421.8
		(23.93)		(55.58)	(1.40)	(19.09)	(100.00)
1991-92	1081976	338.9	1360835	628.2	28.7	275.2	1271.0
		(26.66)		(49.43)	(2.26)	(21.65)	(100.00)
1992-93	1005034	314.8	1525457	703.9	29.2	294.9	1342.8
		(23.44)		(52.42)	(2.18)	(21.96)	(100.00)
1993-94	1063680	332.6	1599269	737	24.6	307.2	1401.4
		(23.73)		(52.59)	(1.76)	(21.92)	(100.00)
1994-95	0	353.2	0	708.9	33.2	304.8	1400.1
		(25.23)		(50.63)	(2.37)	(21.77)	(100.00)
1995-96	0	403.9	0	655.2	29.2	319.1	1407.4
		(28.70)		(46.55)	(2.08)	(22.67)	(100.00)
1996-97	0	390.4	0	694.8	23.8	359.9	1691.1
		(26.58)		(47.30)	(1.62)	(24.50)	(100.00)
1998-99	0	335.6	0	891.5	16.3	325.2	1568.6
		(21.39)		(56.83)	(1.04)	(20.74)	(100.00)

Contd...

Table 5.6 : (Contd....)

Period	*Hessian*		*Sacking*		*CBC*	*Others*	*Total*
	000' Meter	*Ton*	*000' Meter*	*Ton*	*000' Meter*	*Ton*	*Ton*
1999-2000	0	313.9	0	822.6	6.7	305.8	1449.0
		(21.66)		(56.77)	(0.46)	(21.11)	(100.00)
Average	484107.27	348.74	704581.09	751.12	23.70	309.90	1433.45
SD	530730.79	27.56	775661.35	88.26	7.36	36.02	109.28
CV	109.63	7.90	110.09	11.75	31.05	11.62	7.62
AGR	102.29	108.57	97.06	87.08	447.76	84.63	92.88
Annual Growth Rate		1.52		15.39	-1.51	8.42	22.83

Source : Annual Jute Goods Statistics 2000, Bangladesh.

Table indicates that Annual production was 1433.45, SD was 109.28, CV was 7.62, Average 92.88.

Figures in the parentheses indicate percentage to total.

Table 5.7 : Export Quantity in Tone (BJMC & BJMA)

Period	*Hessian*	*Sacking*	*CBC*	*Carpet*	*Others*	*Total*
1989-90	195787	209838	69018	514	11687	486844
	(40.22)	(43.10)	(14.18)	(0.10)	(2.40)	(100.00)
1990-91	131097	122128	59187	465	11299	324176
	(40.44)	(37.67)	(18.26)	(0.14)	(3.49)	(100.00)
1991-92	132333	197905	58815	703	17988	407744
	(32.45)	(48.54)	(14.42)	(4.42)	(4.42)	(100.00)
1992-93	123268	215260	48190	994	21026	408738
	(30.16)	(52.66)	(11.79)	(0.24)	(5.15)	(100.00)
1993-94	133462	158488	57156	1755	24552	375413
	(35.55)	(42.22)	(15.22)	(0.47)	(6.54)	(100.00)
1994-95	115037	188590	48135	1539	30890	384191
	(29.94)	(49.09)	(12.53)	(0.40)	(8.04)	(100.00)
1995-96	98173	167704	44591	1857	34258	346583
	(28.33)	(48.38)	(12.87)	(0.54)	(9.88)	(100.00)
1996-97	65512	147276	33332	1660	36132	283912
	(23.07)	(51.87)	(11.75)	(0.58)	(12.73)	(100.00)
1997-98	77556	88824	35307	910	38379	240976
	(32.18)	(36.86)	(14.65)	(0.38)	(15.93)	(100.00)
1998-99	105666	137782	25696	710	32959	302813
	(34.89)	(45.50)	(8.50)	(0.23)	(10.88)	(100.00)
1999-2000	85721	122782	35473	465	24690	269131
	(31.85)	(45.62)	(13.18)	(0.17)	(9.18)	(100.00)
BJMC	70449	98254	27085	284	5521	201593
	(34.95)	(48.74)	(13.44)	(0.14)	(2.73)	(100.00)
BJMA	15272	24528	8388	181	19169	67538
	(22.61)	(36.32)	(12.42)	(0.27)	(28.38)	(100.00)
Average	103794.85	144566.08	42336.38	925.92	23734.62	315357.85
SD	41945.19	52229.44	16171.17	563.76	10036.45	103490.19
CV	40.41	36.13	38.20	60.89	42.29	32.82
AGR	1282.00	855.50	822.82	283.98	60.97	720.84
Annual Growth Rate	-8792.6	-7515.08	-3768.13	25.36	2296.69	720.84

Source : Annual Jute Goods Statistics 2000, Bangladesh.

Table 5.7 shows that annual average export in quantity was 315357.85, SD was 103490.19, CV was 32.82, ARG 720.84. Figures within parentheses indicate percentage to total.

Table 5.8 : Export Value (BJMC & BJMA) (in Million Taka)

Period	*Hessian*	*Sacking*	*CBC*	*Carpet*	*Others*	*Total*
1989-90	4669.2	2930.6	1596.8	47	299	9542.6
	(48.93)	(30.71)	(16.73)	(0.50)	(3.13)	(100.00)
1990-91	3563.7	1865.4	1390.9	46	347.5	7213.5
	(49.40)	(25.86)	(19.28)	(0.64)	(4.82)	(100.00)
1991-92	3448.7	3066	1536.9	67.9	587.8	8707.3
	(39.61)	(35.21)	(17.65)	(0.78)	(6.75)	(100.00)
1992-93	2758.4	3302.6	1269.5	94	614.4	8038.9
	(34.31)	(41.08)	(15.79)	(1.17)	(7.65)	(100.00)
1993-94	3020.9	2512.4	1435.6	161.6	633	7763.5
	(38.92)	(32.36)	(18.49)	(2.08)	(8.15)	(100.00)
1994-95	3000.4	2961.7	1370.4	153.6	820.5	8306.6
	(36.12)	(35.65)	(16.50)	(1.85)	(9.87)	(100.00)
1995-96	3279.1	3135.1	1464.5	190.1	977.5	9046.3
	(36.25)	(34.66)	(16.19)	(2.10)	(10.80)	(100.00)
1996-97	2477.2	3695.2	1147.6	179.1	1129.2	8628.3
	(28.70)	(42.83)	(13.30)	(2.08)	(13.09)	(100.00)
1997-98	2452.1	2135.4	1202.8	106	925.6	6821.9
	(35.94)	(31.30)	(17.63)	(1.55)	(13.58)	(100.00)
1998-99	3208.4	2729.6	885.6	84.8	790.3	7698.7
	(41.67)	(35.46)	(11.50)	(1.10)	(10.27)	(100.00)
1999-2000	2926	2484	1215.1	48	602.8	7275.9
	(40.21)	(34.14)	(16.70)	(0.66)	(8.29)	(100.00)
BJMC	2343.7	1981	939.3	35.9	100.2	5400.1
	(43.40)	(36.68)	(17.39)	(0.66)	(1.87)	(100.00)
BJMA	582.3	503	275.8	12.1	502.6	1875.8
	(31.20)	(26.82)	(14.70)	(0.65)	(26.69)	(100.00)
Average	2902.32	2561.69	1210.06	94.32	640.80	7409.18
SD	886.50	784.05	338.48	56.99	278.30	1898.78
CV	30.54	30.61	27.97	60.42	43.43	25.63
AGR	801.85	582.62	578.97	388.43	59.49	508.72
Annual Growth Rate	-122.10	-1.46	-46.79	-4.30	51.62	-114.43

Source : Annual Jute Goods Statistics 2000, Bangladesh.

Table 5.8 : shows that annual average export value was million Tk. 7409.18, SD was 1898-78 CV was 25.63, ARG 508.72. Figures within the parentheses indicate percentage to total.

Table 5.9 : Internal Consumption in Quantity (BJMC & BJMA) (In Tone)

Period	*Hessian*	*Sacking*	*CBC*	*Carpet*	*Others*	*Total*
Period	Hessian	Sacking	CBC	Carpet	Others	Total
1989-90	18104	32022	663	144	1013	51946
	(34.85)	(61.64)	(1.28)	(0.28)	(1.95)	(100.00)
1990-91	7114	42010	562	112	2281	52079
	(13.66)	(80.66)	(1.08)	(0.22)	(4.38)	(100.00)
1991-92	7260	49261	980	133	2586	60220
	(12.06)	(81.80)	(1.63)	(0.22)	(4.29)	(100.00)
1992-93	5269	41958	890	165	3838	52120
	(10.11)	(80.50)	(1.71)	(0.32)	(7.36)	(100.00)
1993-94	6332	44014	1621	796	5358	58121
	(10.89)	(75.73)	(2.79)	(1.37)	(9.22)	(100.00)
1994-95	5394	52880	1541	958	5501	66274
	(8.14)	(79.79)	(2.33)	(1.44)	(8.30)	(100.00)
1995-96	5674	55166	1000	1685	7615	71140
	(7.98)	(77.55)	(1.41)	(2.36)	(10.70)	(100.00)
1996-97	6953	61915	1292	1499	7853	79512
	(8.74)	(77.87)	(1.62)	(1.89)	(9.88)	(100.00)
1997-98	8421	12189	1570	1537	7848	31565
	(26.68)	(38.62)	(4.97)	(4.87)	(24.86)	(100.00)
1998-99	4809	68469	1427	1094	4884	80683
	(5.96)	(84.86)	(1.77)	(1.36)	(6.05)	(100.00)
1999-2000	4458	68100	1147	1004	4992	79701
	(5.59)	(85.44)	(1.45)	(1.26)	(6.26)	(100.00)
BJMC	1654	39259	634	132	1352	43031
	(3.84)	(91.23)	(1.47)	(0.32)	(3.14)	(100.00)
BJMA	2804	28841	513	872	3640	36670
	(7.65)	(78.65)	(1.40)	(2.38)	(9.92)	(100.00)
Average	6480.46	45852.62	1064.62	779.31	4520.08	58697.08
SD	3793.93	15449.41	385.03	566.07	2255.11	15620.73
CV	58.54	33.69	36.17	72.64	49.89	26.61
AGR	645.65	111.03	129.24	16.51	27.83	141.66
Annual Growth Rate	-647.79	2055.24	71.21	145.43	512.55	2136.63

Source : Annual Jute Goods Statistics 2000, Bangladesh.

Table 5.9 : shows that annul average internal consumption in tone was 58697.08, SD was 15620.73, CV was 26.61, ARG 141.66. Figures within the parentheses indicate percentage to total.

Table 5.10 : Internal Consumption Value (in Million Taka) of BJMC & BJMA

Period	*Hessian*	*Sacking*	*CBC*	*Carpet*	*Others*	*Total*
1989-90	485	589.1	14.1	12.7	21.1	1122
	(43.23)	(52.50)	(1.26)	(1.13)	(1.88)	(100.00)
1990-91	231.7	948.8	13.4	10.2	40.6	1244.7
	(18.61)	(76.23)	(1.08)	(0.82)	(3.26)	(100.00)
1991-92	195.9	1121.6	24.3	11.4	77.5	1430.7
	(13.68)	(78.40)	(1.70)	(0.80)	(5.42)	(100.00)
1992-93	131.6	802.2	22.8	14.3	83.6	1054.5
	(12.48)	(76.07)	(2.16)	(1.36)	(7.93)	(100.00)
1993-94	135.8	877.7	39.1	79.7	86.3	1218.6
	(11.15)	(72.03)	(3.21)	(6.54)	(7.07)	(100.00)
1994-95	128.7	1100.6	40.8	84.3	117.4	1471.8
	(8.74)	(74.78)	(2.77)	(5.73)	(7.98)	(100.00)
1995-96	159.6	1239.2	28.5	151.6	181.1	1760
	(9.07)	(70.41)	(1.62)	(8.61)	(10.29)	(100.00)
1996-97	227.1	1570.2	42.6	134.1	219.2	2193.2
	(10.35)	(71.59)	(1.95)	(6.11)	(10.00)	(100.00)
1997-98	248.3	1544.5	47.9	122	183.2	2145.9
	(11.57)	(71.97)	(2.23)	(5.69)	(8.58)	(100.00)
1998-99	128.8	1208.7	40.3	96.6	112.3	1586.7
	(8.12)	(76.18)	(2.54)	(6.08)	(7.08)	(100.00)
1999-2000	130.2	1349.1	38	76.8	125.5	1719.6
	(7.57)	(78.45)	(2.21)	(4.47)	(7.30)	(100.00)
BJMC	47.8	803.8	21.1	11.1	40.3	924.1
	(5.17)	(86.98)	(2.30)	(1.20)	(4.36)	(100.00)
BJMA	82.4	545.3	16.9	65.7	85.2	795.5
	(10.36)	(68.55)	(2.12)	(8.26)	(10.71)	(100.00)
Average	179.45	1053.91	29.98	66.96	105.64	1435.95
SD	104.62	316.49	11.48	49.06	57.44	419.43
CV	58.30	30.03	38.30	73.26	54.38	29.21
AGR	588.59	108.03	83.43	19.33	24.77	141.04
Annual Growth Rate	-1649	72.78	2.97	11.90	1356	141.04

Source : Annual Jute Goods Statistics 2000, Bangladesh.

Table 5.10 : focuses that annul average value of internal consumption in million Tk. was 1435.95, SD was 419.43, CV was 29.21, ARG 141.04. Figures within the parentheses indicate percentage to total.

Table 5.11 : Weaving Production in India

Period	*Hessian*		*Sacking*		*CBC*		*Carpet*		*Others*	*Total*
	000' Meter	*Ton*	*000' Meter*	*Ton*	*000'Meter*	*000'Ton*	*Ton*		*Ton*	*Ton*
1989-90	160385	46975	133065	59139	13692.0	10854	445		2079.0	119492.0
		(39.32)		(49.49)		(9.08)	(0.37)		(1.74)	(100.00)
1990-91	219252	57100	200373	90257	14823	8813	397		3016.0	159583.0
		(35.78)		(56.56)		(5.52)	(0.25)		(1.89)	(100.00)
1991-92	175052	48618	95900	42622	10103	6406	393		5034.0	103073.0
		(47.17)		(41.35)		(6.22)	(0.38)		(4.88)	(100.00)
1992-93	122053	34750	52042	24206	17038	11363	563		6580.0	77462.0
		(44.86)		(31.25)		(14.67)	(0.73)		(8.49)	(100.00)
1993-94	0	10372	0	41590	0	1584	870		3422.0	57838.0
		(17.93)		(71.91)		(2.74)	(1.50)		(5.92)	(100.00)
1994-95	0	7732	0	11272	0	1425	924		3539.0	24892.0
		(31.06)		(45.28)		(5.72)	(3.71)		(14.23)	(100.00)
1995-96	0	5875	0	10596	0	2081	911		3476.0	22939.0
		(25.61)		(46.19)		(9.07)	(3.97)		(15.15)	(100.00)
1996-97	0	31559	0	12853	0	2225	797		3444.0	50878.0
		(62.03)		(25.26)		(4.37)	(1.57)		(6.77)	(100.00)
1997-98	0	32531	0	70938	0	5192	1387		6919.0	116967.0
		(27.80)		(60.65)		(4.44)	(1.19)		(5.92)	(100.00)

Contd....

Table 5.11 : (Contd....)

Period	Hessian		Sacking		CBC	Carpet		Others	Total
	000' Meter	Ton	000' Meter	Ton	000'Meter	000'Ton	Ton	Ton	Ton
1998-99	0	11078	0	59841	0	13616	944	7802.0	93281.0
		(11.88)		(64.15)		(14.60)	(1.01)	(8.36)	(100.00)
1999-2000	0	17887	0	46443	0	6402	845	5249.0	76826.0
		(23.29)		(60.45)		(8.33)	(1.10)	(6.83)	(100.00)
BJMC	0	14885	0	41309	0	5188	608	1758.0	63748.0
		(23.35)		(64.80		(8.14)	(0.95)	(2.76)	(100.00)
BJMA	0	3002	0	5134	0	1214	237	3491.0	13078.0
		(22.95)		(39.26)		(9.28)	(1.81)	(26.69)	(100.00)
Average	52057.08	24797.23	37029.23	39707.69	4281.23	5874.08	717.00	4293.00	75389.00
SD	80434.26	17483.96	63206.43	25094.21	6570.80	4044.17	300.57	1802.13	41215.45
CV	154.51	70.51	170.69	63.20	153.48	68.85	41.92	41.98	54.67
AGR	131.41	1564.79	255.69	1151.91	80.36	894.07	187.76	59.55	913.69
Annual Growth Rate		-3533.35		-1399.05		-222.45	69.81	313.0	-4772.03

Source : Annual Jute Goods Statistics 2000, Bangladesh.

Figures within the parentheses indicate percentage to total.

Table 5.11 : Focuses that Annual Average weaving production India, 75389.00, SD was 41215.45, CV was 54.67, ARG 913.69.

Table 5.12 : Manpower Position of Enterprises (As on 30-06-2000)

Enterprises	*1989-90*	*1990-91*	*1991-92*	*1992-93*	*1993-94*	*1994-95*	*1995-96*	*1996-97*	*1997-98*	*1998-99*	*1999-00*	*Total*	*Annual Growth Rate*
BJMC	2475	3889	3779	10143	71116	1119	27834	100069	52237	17237	331	69935	2955.78
	(56.46)	(61.98)	(58.80)	(59.36)	(67.64)	(100.00)	(52.00)	62.63)	(80.69)	(64.17)	(14.18)	(74.29)	
BJMA	1909	2386	2648	6943	34026	0	25688	59714	12502	9696	2003	24201	1422.51
	(43.54)	(38.02)	(41.20)	(40.64)	(32.36)	(0.00)	(48.00)	(37.37)	(19.31)	(35.83)	(85.82)	(25.71)	
Total	4384	6275	6427	17086	105142	1119	53522	159783	64739	27063	2334	94136	4378.29
(June 2000)	(100.00)	(100.00)	(100.00)	(100.00)	(100.00)	(100.00)	(100.00)	(100.00)	(100.00)	(100.00)	(100.00)	(100.00)	

Source : Annual Jute Goods Statistics 2000, Bangladesh.

Figures within the parentheses indicate percentage to total.

Table 5.13 : Statement of Production, Export & Internal Consumption and Stock of Jute Goods of (BJMC & (BJMA) (in '000' Tone)

Period	*Opening Stock*	*Baled Prod.*	*Total Export & Internal Conusmption*	*Actual*	*Theoretical (A+B-C) Differences Between Theoretical and Actual Stock*
1989-90	136	515	540	119	111
1990-91	119	426	387	160	158
1991-92	160	409	468	103	101
1992-93	103	433	460	77	76
1993-94	77	416	438	58	55
1994-95	58	416	451	25	23
1995-96	25	412	418	23	19
1996-97	23	395	364	51	54
1997-98	51	398	329	117	120
1998-99	117	358	384	93	91
1999-2000	97	333	349	77	81
BJMC	82	231	245	64	68
BJMA	15	102	104	13	13
Average	81.77	372.62	379.77	75.38	74.62
SD	43.89	99.86	106.65	41.21	40.90
CV	53.67	26.80	28.08	54.66	54.82
AGR	906.67	504.90	519.23	915.38	853.85
Annual Growth Rate	-6.75	-11.77	-14.51	-4.75	-4.00

Source : Annual Jute Goods Statistics 2000, Bangladesh.

Annual Export value was 379.77, SD was 106.65, CV was 28.08, Average 519.

Performance of Trade Union in Public Sector Jute Mills

This section summarises the socio-economic background of trade union members' opinions regarding attitude of management, implications of Trade Union Acts, cause of disputes, nature of disputes, reconciliation mechanism, productivity, financial and non-financial benefit, role of trade union, conflicts with management and other functional aspects of trade union. The socio-economic background of the union members is represented in tables 5.16 to 5.41. We have collected from two samples (sample 1 and sample 2). Sample 1 represents Jessore Jute Industries Ltd. (JJI) and sample 2 represents the Rajshahi Jute Mills Ltd. (RJM). Both the mills are nationalised jute mills. The samples are taken with simple random sampling without replacement. We have represented the data in tabular forms and χ^2 (chi-square) distribution is used to test the significance of the population variation through confidence intervals. In a nutshell we can say that χ^2 distribution is used to judge whether the sample has been drawn from a normal population with mean μ and standard deviation σ or not. χ^2 distribution is defined as :

$$\chi^2 = \frac{N\sigma_x^2}{\sigma^2} = \frac{(\chi_1 - \overline{\chi}_1)^2 + (\chi_2 - \overline{\chi}_2)^2 + \cdots + (\chi_n - \overline{\chi}_n)^2}{\sigma^2}$$

with (n –1) degrees of freedom. Some of the results of Chi-square tests are shown to the bottom of the respective tables representing the data of our opinion survey.

Out of 80 workers of sample 1, 22 workers are in age group of 46-50 years. 20 members belong to the age group of 41-45 years and 11 workers of 31-35 years of age group. 9 members are 36-40 and another 9 members are 51-55 age group. Only 6 members are within 21-25 and 3 members are in age group of 26-30 years. While in sample 2, 18 workers are in age group of 36-40, 15 workers are in 41-45 age groups. 10 workers belong to the age group of 26-30, 31-35, 46-50, and 51-55 respectively. 5 workers belong to 21-25 and 2 workers belong to 56-60 age group.

Table 5.15 show that 38 workers are up to class V standard, 3 members or workers are up to class X standard onl; 10 workers are S.S.C or above and only graduate is available in

Table 5.14 : Classification of Workers on the Basis of the Age

Class Interval (Years)	*Sample 1*		*Sample 2*		*Combined*
	Frequency	*%*	*Frequency*	*%*	*(%)*
21-25	06	7.5	05	6.25	6.88
26-30	03	3.75	10	12.5	8.13
31-35	11	13.45	10	12.5	13.13
36-40	09	11.25	18	22.5	16.87
41-45	20	25.00	15	18.7	21.87
46-50	22	27.00	10	12.5	20.00
51-55	09	11.25	10	12.5	11.87
56-60			02	2.5	1.25
Total	80	100	80	100	100

Source : Field survey.

Sample 1	**Sample 2**
$\chi^2 = 26.16$	$\chi^2 = 17.8$
d.f. = 6	d.f0. = 7
μ = 41.4 years	μ = 39.75 years
σ = 8.45 years	σ = 9 years
Coefficient of variance = 20.4%	Coefficient of variance = 22.6%

Tested homogeneity

Table 5.15 : Classification on the Basis of Education

Class Internal (Years)	*Sample 1*		*Sample 2*		*Combined*
	Frequency	*%*	*Frequency*	*%*	*(%)*
Up to class V	38	47.5	43	53.75	50.63
Up to class X	31	38.75	29	36.25	37.50
S.S.C. & above	10	12.5	08	10.00	11.25
B.A.	01	1.25			0.62
Total	80	100	80	100	100

Source : Field survey.

Sample 1	**Sample 2**
$\chi^2 = 14.806$	$\chi^2 = 23.371$
d.f. = 3	d.f. =2
Not equal	Equal

Tested homogeneity

sample 1. While in sample 2, 43 workers have education up to class V, 29 workers have up to X, 8 workers have up to S.S.C. and above only.

Experience level shows that 22 workers have 25 years or more than that experiences 19 workers have 20-24 years and 18 workers have 15-19 years of experience. 13 worker have 5-9 years experience and another 8 workers have 10-14 years experience in sample 1. In sample 2, 6 workers have 20-24 and 19 workers have 15-19 years of experience. 15 workers have 15 years and above and 12 workers have 10-14 years experience and another 8 workers have 5-9 years experience.

Table 5.16 : Parental Occupation

Types of occupation	*Sample 1*		*Sample 2*		*Combined (%)*
	Frequency	*%*	*Frequency*	*%*	
Agriculture	44	55	37	46.25	50.63
Business	20	25	26	32.5	28.75
Service	16	20	17	21.25	20.62
Total	80	100	80	100	100

Source : Field survey.

Sample 1	**Sample 2**
$\chi^2 = 17.269$	$\chi^2 = 7.580$
d.f. = 2	d.f. = 2
Not equal	Equal

Tested homogeneity

Tab $\chi^2_{.05} = 7.82$

In sample 1 parental occupation reveals that 44 workers' parents are engaged in agriculture, another 20 are engaged in business and 16 are engaged in service respectively. While in sample 2, 37 workers parents are engaged in agriculture, 26 engaged in business and 17 are engaged in service respectively. So the majority of working class came from agricultural background.

Out of 80 workers of sample only 38 workers joined the trade union after joining the mill, 22 joined after few days while

20 workers joined the trade union after few weeks. In case of second sample 34 workers joined the trade union after joining the mill, 35 joined the trade union after few days and 11 joined the trade union after few weeks.

Table 5.17 : Workers Encouragement to Join the Trade Union

	Sample 1		*Sample 2*		*Combined*
	Frequency	*%*	*Frequency*	*%*	*(%)*
Influenced by colleagues	36	45	39	48.75	46.88
As trade union works for the workers	18	22.5	16	20	21.25
Influenced by trade union leaders	17	21.25	15	18.75	20.00
As appointed by trade union leaders	09	11.25	10	12.5	11.87
Total	80	100	80	100	100

Source : Field survey.

Sample 1	**Sample 2**
$\chi^2 = 19.5$	$\chi^2 = 25.10$
d.f. = 3	d.f. =3
Not equal	Equal

Tested homogeneity

Tab $\chi^2_{.05} = 9.84$

Regarding the encouragement of joining the trade union activity 36 workers narrated that they were influenced by their colleagues, 18 workers were motivated as trade union works for the working class, and 17 were influenced by trade union leaders while 9 members got impetus because of their appointment by trade union leaders. Wile in sample 2, 39 workers said that they were influenced by their colleagues, 16 workers opined that trade union workers motivated them. 15 workers were influenced by trade union leaders and 10 members got incentive as they were appointed by trade union leaders.

According tŏ table 5.18, as the sample shows, 28 workers view that their role is to some extent satisfactory, 23 workers opined it as satisfactory while another 23 workers treat it as

Table 5.18 : Role of Trade Union

	Sample 1		*Sample 2*		*Combined*
	Frequency	*%*	*Frequency*	*%*	*(%)*
Their role is alright	23	28.75	10	12.5	20.63
Their role is not satisfactory	23	28.75	30	37.5	33.12
Some times they do Maximum for management	06	7.5	10	12.5	10.00
Their role is to some extent Satisfactory	28	35	30	37.5	36.25
Total	80	100	80	100	100

Source : Field survey.

Sample 1	**Sample 2**
$\chi^2 = 13.9$	$\chi^2 = 20$
d.f. = 3	d.f. = 3
Not equal	Equal

Tested homogeneity

Tab $\chi^2_{.05} = 9.84$

partial satisfactory and the rest 6 workers feel that they do maximum for management. According to second sample 30 workers said that the role of trade union is not satisfactory and another 30 workers opined that their role is to some extent satisfactory while 10 workers think that they do maximum for management.

Table 5.19 : Management's Attitude to Workers

	Sample 1		*Sample 2*		*Combined*
	Frequency	*%*	*Frequency*	*%*	*(%)*
No body is sympathetic to our demand	07	8.75	12	15	11.88
Management is not that much sympathetic	27	33.75	22	27.5	30.62
To some extent they are sympathetic	39	48.75	39	48.75	48.75
Management is not considerate as they gat union support	07	8.75	07	8.75	8.75
Total	80	100	80	100	100

Source : Field survey.

Management's attitude to workers is reflected in the sense that 39 workers think that management is, to some extent sympathetic, 27 workers think that management is not that much sympathetic, 7 workers opine that nobody is sympathetic to the workers' demand while other 7 workers express their grumble that management is not considerate as they get union leaders support. In sample 2, 39 workers think that management is to some extant sympathetic, 22 workers say that management is not that much sympathetic, 12 workers express their opinion that nobody is sympathetic to workers demand but 7 workers say that management is now considerate as they get union leaders support.

Table 5.20 : Ways of Presenting Demands

Ways of Placing Demands	*Sample 1*		*Sample 2*		*Combined*
	Frequency	*%*	*Frequency*	*%*	*(%)*
We place our demand through our union leaders	36	45	29	36.25	40.63
Through direct discussion	19	23.75	20	25	24.37
Through informal communication	11	13.45	16	20	23.12
Written ultimatum	14	17.5	15	18.75	18.13
Total	80	100	80	100	100

Source : Field survey.

Sample 1
$\chi^2 = 18.7$
d.f. = 3
Not equal

Sample 2
$\chi^2 = 6.1$
d.f. = 3
Equal

Tested homogeneity
Tab $\chi^2_{.05} = 9.84$

On the question of presenting demands sample 1 shows that 36 workers assert that they place their demand through their union leaders. 19 workers desire to follow direct discussion, 14 wants to follow written ultimatum and 11 workers exert pressure through informal communication. In sample 2, 29 workers say that they place their demand through their union leaders, 20 workers like direct discussion, 16 workers prefer informal discussion while 15 workers are in favour of giving written ultimatum.

When the answer is negative to the workers demand 33 workers blame for unsympathetic attitude of management, 24 workers tell about misunderstanding, while 23 workers express their sorrow regarding arrogant attitude of management. In sample 2, 32 workers blame unsympathetic attitude to management, 29 workers locate the issue of misunderstanding and 19 workers blame the arrogant attitude of management.

Table 5.21 : Negotiation Process

	Sample 1		*Sample 2*		*Combined*
	Frequency	*%*	*Frequency*	*%*	*(%)*
Through C.B.A. leaders	34	42.5	23	28.75	35.63
Through departmental head	14	17.5	06	7.5	12.50
Direct approach	08	10.0	17	21.25	15.62
Through strike	16	20.0	16	20	20.00
Through written ultimatum	08	10.0	18	22.5	16.25
Total	80	100	80	100	100

Source : Field survey.

Sample 1	**Sample 2**
$\chi^2 = 28.5$	$\chi^2 = 9.625$
d.f. = 4	d.f. = 4
Not equal	Equal

Tested homogeneity

Tab $\chi^2_{.05} = 11.67$

According to the opinions of the respondents of sample 1, negotiation process comprises five techniques. 34 workers think that it should be done through C.B.A. leader, 16 workers describe it through strike, 14 workers feel that it is made through departmental head, 8 workers recommend for direct approach while 8 workers relate it to written ultimatum. In sample 2, 23 workers say that it should be done through CBA leaders, 18 workers prefer going through written ultimatum, 17 respondents opt for direct approach, 16 workers choose the way of going through strike while 6 workers like to proceed through departmental head.

Table 5.22 : Volume of Disputes in Last Five Years

	Sample 1		*Sample 2*		*Combined*
	Frequency	*%*	*Frequency*	*%*	*(%)*
Can not say the exact number	37	46.25	36	45	45.63
So many disputes take place	22	27.5	25	31.25	29.37
It took several times	21	26.25	19	23.75	25.00
Total	80	100	80	100	100

Source : Field survey.

Sample 1	**Sample 2**
$\chi^2 = 6.06$	$\chi^2 = 5.62$
d.f. = 2	d.f. = 2
Not equal	Equal

Tested homogeneity

Tab $\chi^2_{.05} = 7.82$

Regarding the volume of disputes it is found, as sample 1 shows, that 37 workers cannot say the exact number, 22 workers follow that so many disputes take place and the rest 21 workers view that disputes take place several times. So far the answer of second sample is almost similar to sample 1. 36 workers cannot say the exact number, 25 respondents say that so many disputes take place and 19 say that disputes take place several times.

Sample 1 shows 32 workers settle disputes through negotiation with C.B.A. and another 32 workers recommend for direct negotiation while 16 workers believe in mounting pressure through strikes. But in sample 2, 36 workers like to face disputes through negotiation with CBA, 37 workers prefer amicable settlement between management and union leaders but 17 workers believe in making pressure through strikes.

Table 5.23 : Settlement of Disputes

	Sample 1		*Sample 2*		*Combined*
	Frequency	*%*	*Frequency*	*%*	*(%)*
Through negotiation (amicable settlement between union leaders & management)	32	40	27	33.75	36.88
Negotiation through C.B.A.	32	40	36	45	42.50
By giving pressure through strikes	16	20	17	21.25	20.62
Total	80	100	80	100	100

Source : Field survey.

Sample 1	**Sample 2**
$\chi^2 = 6.457$	$\chi^2 = 6.829$
d.f. = 2	d.f. = 2
Not equal	Equal

Test homogeneity

Tab $\chi^2_{.05} = 7.82$

Table 5.24 : Causes of Disputes

	Sample 1		*Sample 2*		*Combined*
	Frequency	*%*	*Frequency*	*%*	*(%)*
Main causes are wages and other benefits	41	51.25	27	33.75	42.50
For financial and non-financial benefit	17	21.25	37	33.75	33.75
Main cause are improved scale and other facilities	22	27.5	16	20	23.75
Total	80	100	80	100	100

Source : Field survey.

According to opinions of the 41 workers in sample 1, main causes of disputes are wages and other benefits, 22 workers say that main causes are improved scale and other facilities, 17 workers think financial and non-financial benefit are the major causes. In sample 2, 37 workers highlight the causes of disputes are financial and non-financial benefit, 27 workers think that the main causes of disputes are wages and other benefits while 16 locate the causes of disputes are improved scale and other facilities.

On the question of settling disputes sample 1 shows, 33 workers take management as sometimes cordial, another 30 take it as partially cordial while 9 workers say management remains silent in settling the disputes. According to sample 2, 31 workers take the attitude of management cordial in setting disputes, 24 say sometimes management is cordial and say management remains silent in setting disputes.

Table 5.25 : Nature of Workers Co-operation

	Sample 1		*Sample 2*		*Combined*
	Frequency	*%*	*Frequency*	*%*	*(%)*
Cooperate at the time of necessities	14	17.5	21	26.25	21.8
We go ahead when management seeks our cooperation	34	42.5	29	36.25	39.37
Cooperate without hampering our interest	17	21.25	17	21.25	21.25
Workers cooperate unconditionally	15	18.75	13	16.25	17.50
Total	80	100	80	100	100

Source : Field survey.

Sample 1	**Sample 2**
$\chi^2 = 13.3$	$\chi^2 = 7.0$
d.f. = 3	d.f. = 3
Not equal	Equal

Tested homogeneity

Tab $\chi^2_{.05} = 9.84$

On workers co-operation as sample 1 shows, it is found that 34 workers go ahead when management seeks for their co-operation. 17 workers co-operate without hampering their interest, 15 workers co-operate unconditionally and 14 workers co-operate at the time of necessity of management. In sample 2, 29 workers cooperate when management seeks their cooperation, 21 workers ready to cooperate at the time of necessity of management, 17 workers cooperate without hampering their interest while 13 workers ready to cooperate unconditionally.

On the non co-operation of trade union member according to sample 1, 36 opine that management then contacts C.B.A. leaders; 26 say that management sometimes accepts their demand partially; 18 say that management comes forward and throws some benefits. From sample 2, 39 workers opine that when trade union dose not cooperate then management contacts CBA leaders, 26 workers say that management some times accept their demand and 15 workers tell that management comes forward with some benefits.

Table 5.26 shows in the sample that 29 workers demand for new scale, 21 workers demand for letter housing and other facilities, 17 workers demand new scale considering price level while 13 workers demand higher wages and other benefits. In the second sample we see that out of 80 workers demand higher wage and other benefits, 28 workers demand new scale considering price level, 12 workers demand new scale and another 12 want better housing and other facilities.

According to sample 1, on the question of ratio of fulfilment of demand 34 respondents claim that very few demands are fulfilled, 25 workers view that few demands are accepted by the management, 14 workers think it as negligible while 7 workers take management as aggressive. On the question of ratio of fulfilment of demand we see that 37 workers say that few demand management accepted, 23 workers complain that very few demands are accepted, while 12 workers think it as negligible and another 12 workers consider management as aggressive to the workers demand according to Sample 2.

Table 5.26 : Nature of Demand in Last Five Years

	Sample 1		*Sample 2*		*Combined*
	Frequency	*%*	*Frequency*	*%*	*(%)*
Demanding new scale	29	36.25	12	15	25.62
Higher wage and other benefits	13	16.25	28	25	25.62
For better housing & other facilities	21	26.25	12	15	20.63
Considering price level We demand new scale	17	21.25	28	35	28.13
Total	80	100	80	100	100

Source : Field survey.

Sample 1	**Sample 2**
$\chi^2 = 7$	$\chi^2 = 12.8$
d.f. = 3	d.f. = 3
Not equal	Equal

Tested homogeneity

Tab $\chi^2_{.05} = 9.84$

Table 5.27 : Workers Reaction on Non-fulfillment of Demand

	Sample 1		*Sample 2*		*Combined*
	Frequency	*%*	*Frequency*	*%*	*(%)*
We go for movement	1	38.75	20	25	31.88
We give ultimatum for greater movement	32	40	36	45	42.50
We give demonstration	08	10	13	16.25	13.12
We go for strike	09	11.25	11	13.45	12.50
Total	80	100	80	100	100

Source : Field survey.

Sample 1	**Sample 2**
$\chi^2 = 26.5$	$\chi^2 = 19.3$
d.f. = 3	d.f. = 3
Not equal	Equal

Tested homogeneity

Regarding the reaction on non-fulfilment of demand 32 trade union members or workers view that they give ultimatum for greater movement, 31 assert that they go for movement, 9 workers going for strike and 8 workers believe in demonstration. According to sample 2, 36 trade union workers react on the non-fulfilment of demand and they want to give ultimatum for greater movement, 20 workers want to go for movement, 13 workers are in favour of giving demonstration and while the rest 11 are in favour of going for strike.

Regarding the awareness of trade union acts, wages acts, profit participation act 28 workers have no detailed knowledge, 27 workers have very little knowledge, 16 workers have no idea in detail, 06 workers simply heard about the acts and 03 do not know at all. In sample 2, 30 workers have not detailed knowledge about the acts, 25 workers have very little knowledge, 10 workers have not detailed knowledge, 12 workers hear about the acts while 03 workers do not know at all.

Table 5.28 : Legal Provision Followed in the Mills

	Sample 1		*Sample 2*		*Combined*
	Frequency	*%*	*Frequency*	*%*	*(%)*
To some extent	32	40	33	41.25	40.63
Not always	19	23.75	17	21.25	22.50
Yes, they try to follow	20	25	24	30	27.50
We have no ide about that	09	11.25	06	7.5	9.37
Total	80	100	80	100	100

Source : Field survey.

Regarding the application of legal provisions it is found in sample 1 that 32 respondents say that legal provisions are to some extant followed. 20 workers view that it is followed some how, 19 workers opine that it is not always followed while 09 workers have no idea about that. According to sample 2, 32 trade union workers react on the non-fulfillment of demand and they want to give ultimatum for greater movement, 20 workers want to go for movement, 13 workers are in favour of giving demonstration and while the rest 11 are in favour of going on strike.

Table 5.29 : Impact of Non-compliance with Legal Provisions

	Sample 1		*Sample 2*		*Combined*
	Frequency	*%*	*Frequency*	*%*	*(%)*
Industrial or human relation will be disturbed	27	33.75	26	32.5	33.12
Personnel problem will come up	16	20	23	28.75	24.38
Legal environment will be threatened	12	15	13	16.25	15.63
Labour problem will emerge	25	31.25	18	22.5	26.87
Total	80	100	80	100	100

Source : Field survey.

Non-compliance of legal provision shows that 27 workers take it as disruption of human relation, 25 workers think that labour problem is aggravated, 16 workers think that personal problem will come up while 12 workers treat it as threat to legal environment. According to sample 2, 26 workers say that industrial relation will be disturbed, 23 workers think that labour problem is aggravated, 18 workers say that personnel problem will come up while 13 workers consider it as threat to legal environment.

Regarding the legal provisions workers' suggestions are as follows in sample 1, 29 workers suggest that management should follow the legal provisions 21 workers recommended for looking after the interest of the workers, 17 recommended for minimizing corruption, 13 recommend for management audit to minimize losses. In second sample, workers' suggestion for it are as follows : 31 workers comment that management should look after the interest of the workers, 22 workers comment that management should follow the legal provisions, 18 workers suggest that management should try to minimise corruption while 9 workers opine that management should try to minimize losses through management audit.

Table 5.30 : Workers Cooperation in Management

	Sample 1		*Sample 2*		*Combined*
	Frequency	*%*	*Frequency*	*%*	*(%)*
Cooperate with in management to meet up da to day crises	25	31.25	23	28.75	30.00
Maintain close cooperation with the management	27	33.75	33	41.25	37.50
Always extend my helping hand to the management in resolving any crisis	16	20	14	17.5	18.75
We are ready to held management as much as we can	12	15	10	12.5	13.75
Total	80	100	80	100	100

Source : Field survey.

Regarding the cooperation of trade union member, it is found that 27 workers maintain close cooperation with the management, 25 workers cooperate with management to meet up day to day crisis, and 16 workers help management resolving any sort of crises and the rest 12 are ready to help management as much as they can. In sample 2, 33 workers maintain close cooperation with the management, 23 workers cooperate with the management to meet up day-to-day crisis, 14 workers help management in resolving any sort of crisis while 10 workers help the management as much as they can.

In this table we find little variation regarding the close cooperation with the management. In sample 2, management-worker cooperation is more than the sample 1, which is helpful to maintain industrial peace.

On the question of rising productivity 35 workers in sample 1 recommend for training of workers and proper wage benefits,

Table 5.31 : Workers View Regarding the Increased Productivity

	Sample 1		*Sample 2*		*Combined*
	Frequency	*%*	*Frequency*	*%*	*(%)*
Productivity could be increased by giving training proper wage and after benefits	35	43.75	32	40	41.88
Productivity could be increased by supplying raw materials proper wages and other benefits	32	40	29	36.25	38.12
Careful observation of Management and fairly payment of wage and other benefits	13	16.25	19	23.75	20.00
Total	80	100	80	100	100

Source : Field survey.

Sample 1	**Sample 2**
$\chi^2 = 10.75$	$\chi^2 = 3.517$
d.f. = 2	d.f. = 2
Not equal	Equal

Tested homogeneity

Tab $\chi^2_{.05} = 7.82$

32 workers recommend for supplying more working capital, raw materials, wage and other benefits while 13 workers hold the view that careful observation of management and fair payments of wages and other benefits. According to sample 3, 30 workers recommended for training for workers, proper payment of wages and other benefits, 29 workers for sufficient supply of raw materials, fair payment of wages and benefits, 19 workers have responded for careful observation of management and fair wages payment for rising productivity.

30 workers have focused that the major problems are financial, administrative and related to personnel, 25 workers think that production has gone down due to storage of raw materials and working capital while 25 others tag it to

Table 5.32 : Major Problems of the Mills

	Sample 1		*Sample 2*		*Combined*
	Frequency	*%*	*Frequency*	*%*	*(%)*
Production has gone down due to shortage of raw materials and working capital	25	25	24	30	30.63
Major problem of our mill are financial, administrative and personnel	30	50	22	27	32.50
Production cost is high and it could be minimized by reducing corruption	25	25	34	42.5	36.87
Total	80	100	80	100	100

Source : Field survey.

reducing corruption. In Sample 2, 34 workers have opined that production cost can be minimised by reducing corruption, 24 workers say that production has gone down to shortage of raw materials and working capital and 22 workers have opined that major problems are financial, administrative and personnel.

31 workers consider managements consistent negligence to the problems of the workers which is one of the major cause, 20 workers emphasise administrative and financial problem, 15 workers think that the main causes are corruption, low wages and other benefits while 14 workers the main of the problems are corruption with the indirect help of C.B.A. leaders. In sample 2, 35 workers opined that main caused of the problem are administrative and financial, 20 workers consider management's consistent negligence to the workers' problem is the main cause, 20 workers have identified corruption, low wage and other benefits as the main cause whereas 5 workers have expressed the view that wide scale corruption with the help of CBA leaders is the major cause of the problems of the mill.

Table 5.33 : Problems of the Mills : Major Causes

	Sample 1		*Sample 2*		*Combined*
	Frequency	*%*	*Frequency*	*%*	*(%)*
The major causes are administrative and financial (shortage of raw materials & working capital).	31	38.75	20	25	31.88
Managements consistent negligence to the problems of the worker	20	25	35	43.75	34.37
Wide spread corruption with the indirect help of the CBA leaders	14	17.5	05	6.25	11.87
The main causes are Corruption and low wage and other benefits	15	18.75	20	25	21.88
Total	80	100	80	100	100

Source : Field survey.

Sample 1	**Sample 2**
$\chi^2 = 9.1$	$\chi^2 = 22.5$
d.f. = 3	d.f. = 3
Equal	Not equal

Tested homogeneity

Tab $\chi^2_{.05} = 9.84$

Majority of the workers suggest to minimize corruption in sample 1 and 2. Out of 80 workers, 34 workers view that management look after the problems of the mill and they should try to minimise corruption, 26 workers have opined that management should have to ensure logistic support to run the mill properly and they have to pay wage and other benefit in time and 20 workers think that management should be more sympathetic. According to sample 2, 35 respondents opine that management should look after the problems of the mill and should try to minimise corruption, 30 respondents say that management should try to ensure logistic support to the mill

Table 5.34 : Workers Suggestions for Improvement

	Sample 1		*Sample 2*		*Combined*
	Frequency	*%*	*Frequency*	*%*	*(%)*
Management should look after the problems of the mill management should try to minimize corruption	34	42.5	35	43.75	43.13
Management should have to ensure logistic problems to run the mill properly. They have to pay wage and other benefits in time	26	32.5	30	37.5	35.00
Management should look after workers sympathetically	20	25	15	18.75	21.87
Total	80	100	80	100	100

Source : Field survey.

Sample 1	**Sample 2**
$\chi^2 = 3.739$	$\chi^2 = 8.188$
d.f. = 2	d.f. = 2
Equal	Not equal

Tested homogeneity

Tab $\chi^2_{.05} = 7.82$

and try to pay wages properly while 15 respondents say that management should be more sympathetic to workers.

On the question of adequacy of salary and other benefits 28 workers are dissatisfied, 22 workers think it more rough facility, 19 workers take it as insufficient while 11 workers have grumbling over absolutely minimum wage. In sample 2, 24 workers are dissatisfied regarding inadequacy of salary, 21 workers say that wages are absolutely minimum, 20 workers say that in comparison to market price it is very much insufficient and 15 workers take it insufficient.

On the question of rationality of the payment of expected wages and other benefits it is found that 31 workers feel its necessity for more output, 18 workers relate it to develop personnel relation and other 18 advocate to minimise labour unrest, 13 workers suggest to develop job satisfaction, 26 workers say that management tries to provide workers with expected wages to develop personnel relation, 23 say that to minimise labour unrest, 19 expressed that expected wages is given to get more output while 12 have opined to minimise labour unrest.

Majority of the workers are against golden handshake 41 workers have the negative attitude regarding golden hand shake, 22 workers ready to accept golden handshake with total service benefit for the creation of employment opportunity for the next generation. 17 workers are ready to accept if full benefit is given. In sample 2, 34 workers are not interested to accept golden hand shake, 23 workers are ready to accept if they get full benefit while another 23 passed the opinion that they are ready to leave the job for the next generation if they get full benefit.

Table 5.35 : Nature of Working of C.B.A.

	Sample 1		*Sample 2*		*Combined*
	Frequency	*%*	*Frequency*	*%*	*(%)*
Works nicely	22	27	22	27	27.50
Not up to mark	37	46.25	33	41.25	43.75
We are no satisfied	17	21.25	18	22.5	21.88
They serve the purpose of management	04	5	07	75	6.87
Total	80	100	80	100	100

Source : Field survey.

On the proper functioning of C.B.A. it is found that 37 workers take it not up to the mark, 22 workers think that they work nicely, 17 workers are not satisfied while 04 workers complain that they serve the purpose of management. According to sample 2, 33 workers say not upto the mark, 22 workers say that C.B.A. works nicely, 18 workers are not satisfied while 7 alleged that they serve the purpose of management.

On the conflict among the different trade unions it is found that 38 workers find no conflict within the factory as only one trade union is working, 23 workers find it very rare and 19 workers believe that it occurs sometime. It sample 2, 30 workers say that there is only one trade union, 25 workers find it very rare, 13 and other 12 workers say it happens sometimes.

Table 5.36 : Causes of Conflict Between Different Trade Unions

	Sample 1		*Sample 2*		*Combined*
	Frequency	*%*	*Frequency*	*%*	*(%)*
Due to political reason	28	35	22	27	31.25
Due to leadership conflict	34	42.5	23	28.75	35.63
Due to personality clash	18	22.5	35	43.75	33.12
Total	80	100	80	100	100

Source : Field survey.

Sample 1	**Sample 2**
$\chi^2 = 4.95$	$\chi^2 = 3.87$
d.f. = 2	d.f. = 2
Equal	Equal

Tested non-homogeneity
Tab $\chi^2_{.05} = 7.82$

Causes of conflict reveal that 34 workers find leadership conflict, 28 workers find political reason for this and 18 workers think personality clash. While in sample 2, 35 workers find personality clash as the cause of conflict, 23 workers find leadership conflict and 22 workers think that causes are political.

On the suggestions of removing the conflict 33 workers suggest for minimising politicisation of unions, 22 workers demand early settlement of disputes, 15 workers stress on participation of workers in management and union activities while 10 workers recommend for direct negotiation with workers. According to sample 2, 27 respondents go for minimising politicalisation of unions and another 27 opine for

Table 5.37 : Suggestions for Removing the Conflict

	Sample 1		*Sample 2*		*Combined*
	Frequency	*%*	*Frequency*	*%*	*(%)*
We should try to minimize politicization of unions	33	41.25	27	33.75	37.50
Early settlement of disputes	22	27	27	33.75	30.63
Workers participation in management and union activities	15	18.75	15	18.75	18.75
Direct negotiation with workers	10	12.5	11	13.45	13.12
Total	80	100	80	100	100

Source : Field survey.

Table 5.38 : Workers Suggestions to Management for Development Trade Union

	Sample 1		*Sample 2*		*Combined*
	Frequency	*%*	*Frequency*	*%*	*(%)*
Training for trade union members	13	16.25	23	28.75	22.50
Motivation of trade union members	11	13.45	28	35	24.38
Management should take the trade union as integral part of the enterprise	32	40	09	11.25	25.62
Close co-operation between management and workers	24	30	20	25	27.50
Total	80	100	80	100	100

Source : Field survey.

Sample 1	**Sample 2**
$\chi^2 = 14.5$	$\chi^2 = 9.7$
d.f. = 3	d.f. = 3
Not equal	Not equal

Tested non-homogeneity

Tab $\chi^2_{.05} = 7.82$

early settlement of disputes, 15 workers stress on the participation of workers management and union activities while 11 respondents prefer negotiation.

On the question of making good use of trade union it is available that 32 workers suggest that management should treat the trade union as integral part of the enterprise, 24 workers recommend for close co-operation between management and worker, 13 workers suggest training for trade union members i.e. workers while 11 stress on the motivation of trade union members. In sample 2, 28 workers put stress on motivation, 23 workers opine for training of trade union members, 20 workers have recommended for close cooperation between management and trade unions, 9 workers say that management should consider trade union as integral part of the enterprise.

Table 5.39 : Role of Trade Union

	Sample 1		*Sample 2*		*Combined*
	Frequency	*%*	*Frequency*	*%*	*(%)*
Good, works some how	17	21.25	23	28.75	25.00
So, so ?	22	27	18	22.5	25.00
Not bad	09	11.25	17	21.25	16.25
Satisfactory	32	40	22	27	33.75
Total	80	100	80	100	100

Source : Field survey.

On the role of trade union it is found that 32 workers treat it as satisfactory, 22 workers treat it so-so, and 17 workers consider trade union works somehow. According to sample 2, 23 particpants/workers say good somehow, 22 participant pass the comment as their role is not satisfactory, 18 participants say so-so, while 17 participants take it as not bad.

A Comparative Study of the Two Unions (Sample 1 & 2) Summary

To sum up the overall performance of trade union activities of the 1st Jute mills (Sample-1), favourable aspects of trade union activities are given below in the viewpoint of management. (a) There is no such acute labour dispute, (b) Bipartite negotiation takes place, (c) Disputes reconciled through negotiation, (d) Maximum co-operation attained from C.B.A. (e) Personal relation is good. Demands are fulfilled according to the capacity of the mill, (f) Decentralisation Principle is practiced when it is required, (g) Payment of wages act. 1936 is followed to some extent, (h) Attempts have already been made to follow the Factories Act. 1934, (i) Trade union co-operates in all respect, (j) Employee participation Act 1969 is followed to a limited extent.

Unfavourable aspects are : (a) There is only one trade union but different affiliations of trade union members (Workers) and leaders, (b) Sometimes strikes take place, (c) To combat their demand, sometimes negotiation takes place or pressure is given, (d) Outsiders intervention in union activity some times create problem.

To sum up the over all performance of trade union activities of 2nd Jute mill, favourable aspect of trade union activities are also given below in the view point of management: (a) Direct negotiation takes place with the union leaders, (b) Bipartite negotiation takes place in this mill, (c) Disputes are reconciled through negotiation Management tries to satisfy their demand as they can, (d) Management invites trade union leaders when problems crop up. When management seeks co-operation, workers stretch their hands of co-operation, (e) Management does try to meet up their demands as much as they can (f) When any major issue comes up then participating decision-making takes place, (g) Trade union does not affect productivity always, (h) When workers go on strike, management does contact labour leaders for negotiation, (i) For nationalized mills there is a wage commission. The Commission for all mills under BJMC gives wage and salary according to the prescribed wages, (j) By and large Employee Participating Act. 1969 is followed.

Unfavourable aspects are: (a) Only one trade union in the mill leaders had different affiliations, (b) Disputes take place in different times. It is not possible to say how many disputes took place in last five years due to non-availability of data, (c) Workers' demand for better housing, bonus or other facilities and better working condition, (d) If their demand remains unsatisfied they take up the matter to the management for discussion or sometimes they threaten for greater movement, (e) Due to arrogant attitude of trade union leaders participative decision making or any discussion on any issue sometime fails, (f) co-operative is reciprocal Management-people try to co-operate each other.

Optimistic Views

(a) Union leaders say that their main aim is to help the workers and management to restore industrial peace which is required for smooth running of production and distribution.

(b) Political involvement gives strength to workers by extending legal and moral fight within the legislative body and plant level.

(c) Majority of workers co-operate with the union leaders.

(d) Workers participate spontaneously with the union activity.

(e) Demands are placed to the management after which bipartite negotiation takes place.

(f) Trade union always co-operates with the management to run the factory effectively.

(g) Management is, to some extent, co-operative.

(h) Trade union can play a positive role in increasing the productivity.

(i) Workers feel secured if there is healthy and strong trade union in existence and that looks up morale of the workers as well.

(j) Trade union brings some economic benefits and non-monetary benefits too.

Deterrent Factors

(a) As trade union leaders they are not directly involved in any political party but they take part in national movement.

(b) Trade union leaders maintain political link as political leaders sometimes raise their voice in support of workers and leaders and try to pass bills in favour of them.

(c) Disputes take place for wage, other benefits and illegal retrenchment of workers etc.

(d) Wage is absolutely minimum.

(e) Legal provisions are not followed properly.

(f) As Jute sector is very weak workers cannot demand more, or gear up movement.

(g) Mill is old one, so it requires replacement of machinery. The mill requires more working capital and efficient management and eradication of corruption.

Encouraging Aspects are as Under

First, union leaders world like to serve their colleagues. They want to do something for the working class.

Second, majority of the workers cooperate with the union activity and their participation is spontaneous.

Third, major causes of industrial disputes are regarding wage, bonus and other financial and non-financial benefits.

Fourth, trade union leaders always co-operate with management.

Fifth, present management is more or less co-operative.

Sixth, bipartite negotiation takes place in the mill.

Seventh, if there is healthy trade union there will be little chance of unrest. Thus organised union can play a positive role in increasing productivity.

Eighth, strong trade union ensures high morale of workers.

Ninth, trade union brings socio-economic development of workers. Trade union of Jute mills try to do the same.

Tenth, healthy trade union activity brings job satisfaction.

Finally, management people should be more dedicated to their work. Present management has become more committed that is why mill is earning.

There are Some disliking aspects which are shown below:

First, trade union leaders have political attachment. But they said that they never try to induce it within the factory campus.

Second, union Leaders say that they keep contact with particular political party for their existence. And political leaders help them in crises period.

Third, major causes of disputes are wage, other benefits and illegal retrenchment of workers.

Fourth, present wage structure is poor in comparison to productivity of the workers of Jute mills.

Fifth, trade union Act, 1969 needs some modification.

Sixth, wages Act and Profit participation Act, 1969 are followed somehow.

Finally, union leaders say that basic causes of managerial inefficiency of the Jute mills in Bangladesh are lack of dedication honesty. Sincerity and commitment. After nationalisation every body thought that job is secured. So some people become reckless and corrupt.

Impact of Trade Unions on the Labour Community

The impact of Trade Unions on the labour community is shown in table 5.40, 5.41 and 5.42.

Average equity per employee was Tk. 4.08 lac in Jessore Jute Industries and it was Tk. 1.60 lakh in Rajshahi Jute Mills during 1996-2000.

Table 5.40 : Equity Per Employee (Taka in Lac)

Year	*Total Equity*		*Equity per Employee*			
	JJI	*RJM*	*No. of employee*		*JJI*	*RJM*
1996	14039.68	3605.21	3476	2255	4.04	1.60
1997	14135.83	3605.21	3476	2255	4.07	1.60
1998	14197.06	3605.21	3476	2255	4.08	1.60
1999	14249.22	3605.21	3476	2255	4.10	1.60
2000	14321.78	3605.21	3476	2255	4.12	1.60
Average	14188.71	3605.21	3476	2255	4.08	1.60
SD	96.38	0.00	0.00	0.00	0.03	0.00
CV	0.68	0.00	0.00	0.00	0.68	0.00
Annual Growth Rate	67.76	0	0	0	0.02	-

Source: Annual Jute Goods Statistics 2000, Bangladesh Jute Mills Corporation.

Table 5.41 : Total Assets and Equity Per Employee (Taka in Lac)

Year	*Total assets*		*Equity per Employee*			
	JJI	*RJM*	*No. of employee*		*JJI*	*RJM*
1996	19980.86	6512.14	3476	2255	5.75	2.89
1997	26668.17	7114.38	3476	2255	7.67	3.15
1998	22051.28	8072.72	3476	2255	6.34	3.58
1999	22551.75	1855.96	3476	2255	6.49	0.82
2000	23703.04	2204.35	3476	2255	6.82	0.98
Average	22991.02	5151.91	3476	2255	6.61	2.28
SD	2198.43	2599.39	0.00	0.00	0.63	1.15
CV	9.56	50.45	0.00	0.00	9.56	50.45
Annual Growth Rate	332.79	-1387.4	0	0	0.10	2.28

Source: Annual Jute Goods Statistics 2000, Bangladesh Jute Mills Corporation.

Average equity per employee was 6.61 lac in Jessore Jute Industries and it was Tk. 2.28 lac in Rajshahi Jute Mills during 1996-2000.

Table 5.42 : Loss Accounts of Mills (Taka in Lac)

Year	*Net Loss of Mills*					
	JJI	*RJM*	*No. of employee*		*Net loss perEmployee*	
1996	5.3	155.08	3476	2255	0.00	0.07
1997	393.51	222.26	3476	2255	0.11	0.10
1998	842.9	351.82	3476	2255	0.24	0.16
1999	666.18	323.88	3476	2255	0.19	0.14
2000	776.35	228.18	3476	2255	0.22	0.10
Average	536.85	256.24	3476	2255	0.15	0.11
SD	306.87	71.95	0.00	0.00	0.09	0.03
CV	57.16	28.08	0.00	0.00	57.16	28.08
Annual Growth Rate	181.48	24.78	0	0	0.05	0.11

Source : Annual Jute Goods Statistics 2000, Bangladesh Jute Mills Corporation.

Profit/loss accounts and the rates of growth of profit of the public sector jute mills (nationalised) for a period of six years from 1995 to 2000 have been presented below in tables 5.43 to 5.49.

Table 5.43 : Zone-wise Profit/Loss Accounts of Nationalised Jute Mills of Bangladesh (Taka in thousand)

Zones	*Profit/Loss Accounts for*					
	1995	*1996*	*1997*	*1998*	*1999*	*2000*
Dhaka	2874524	3121690	3584593	4056126	5871431	6546785
Khulna	7022210	7515503	8668293	9974701	9873819	11056063
Chittagong	3600427	3794956	4082590	4592104	5216518	5781994
Adamjee	5307685	6036296	6939984	7876216	8936118	9980883
Total	18804846	20468445	23275460	26499147	29897886	33365725

Source : Compiled from Audited Accounts of Bangladesh Jute Mills Corporation (BJMC) for the year 1995, 1996, 1997, 1998, 1999 and 2000.

The Table 5.43 shows that there has been a steady growth of profit of jute mills under public sector in different zones. This scenario is shown in bar diagram 5.1.

Zone-wise Profit/Loss Account

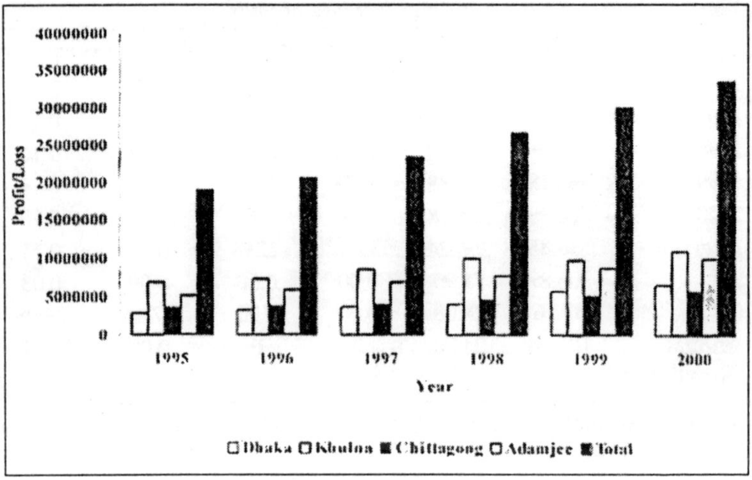

Figure 5.1

The table 5.44 shows that the growth rate of profit of public sector jute mills under Dhaka Zone is quite remarkable (total average 18.54%). The striking improvement in the growth rate of profit took place in 1999 when it showed a 44.75 percent increase in the profit. This is also shown in terms of growth curve (Fig. 5.2) drawn below from the table 5.44.

Table 5.44 : Growth Rate of Jute Mills of Dhaka Zone

(Taka in thousand)

Year	*Profit/Loss*	*Growth Rate*
1995	2874524	-
1996	3121690	8.59
1997	3584593	14.8
1998	4056126	13.1
1999	5871431	44.75
2000	6546785	11.5
Total Average	4342524.833	18.54

Source : Compiled from table 5.43.

Growth Rate of Jute Mills of Dhaka Zone

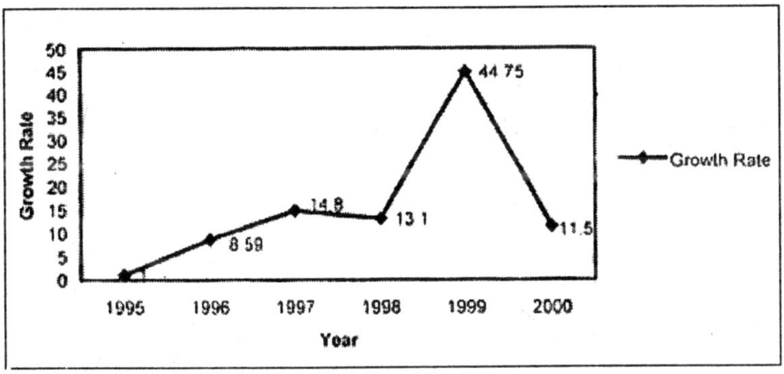

Figure 5.2

Khulna zone's improvement of profit rate is moderate. Over six years (from 1995 to 2000) Khulna zone's average growth rate of profit is 8.45. Both in 1997 and 1998, the growth rates were quite high which were 15.3 and 15.07 percent, respectively. One bad year was 1999 when its profit went down from 9974701 taka to 9873814 taka. Thus the zone experienced a negative growth rate, *i.e.,* 7.02 percent in 1999. The overall position of profit was recorded in 2000. In 2000, the rate of growth of profit went upto 11.9. The growth rate of profit of nationalised jute mills of Khulna zone is shown in Fig. 5.3.

Table 5.45 : Growth Rate of Nationalised Jute Mills of Khulna Zone

(Taka in thousand)

Year	*Profit/Loss*	*Growth Rate*
1995	7022210	-
1996	7515503	7.02
1997	8668293	15.3
1998	9974701	15.07
1999	9873819	7.02
2000	11056063	11.9
Total Average	9018431.5	8.45

Source : Compiled from table 5.43.

Growth Rate of Profit of Nationalised Jute Mills of Khulna Zone

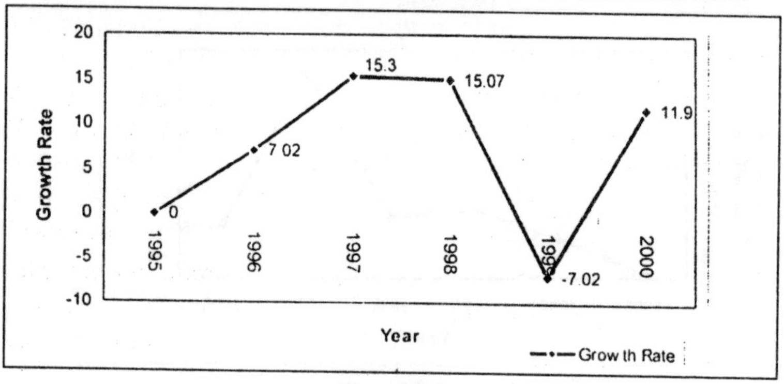

Figure 5.3

Table 5.46 : Growth Rate of Nationalised Jute Mills of Chittagong Zone

(Taka in thousand)

Year	*Profit/Loss*	*Growth Rate*
1995	3600427	-
1996	3794956	5.4
1997	4082590	7.57
1998	4592104	12.4
1999	5216518	13.5
2000	5781994	10.8
Total Average	4511431.5	9.9

Source : Compiled from table 5.43.

Jute mills under Chittagong zone continued to make a profit of 9.9 percent over six years (from 1995 to 2000). In both 1998 and 1999 the Jute mills earned a high profit (4592104 taka and 5216518 taka respectively) and the growth rates of profit were 12.4 and 13.5 percent respectively. It slightly went down from 13.5 percent in 1999 to 10.8 percent in 2000. This scenario is represented in Fig. 5.4.

The condition of nationalised jute mills in Adamjee zone in terms of profit earning is quite satisfactory. The average growth rate of profit over six years from 1995 to 2000 is 13.4 percent. This situation is represented in Fig. 5.5.

Growth Rate of Nationalised Jute mills of Chittagong

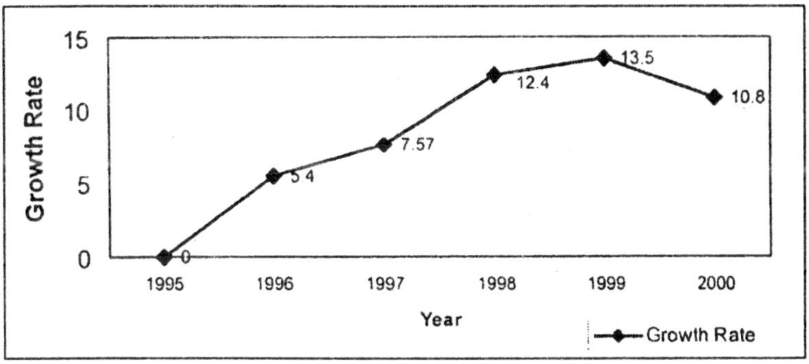

Figure 5.4

Table 5.47 : Growth Rate of Nationalised Jute Mills of Adamjee Zone

(Taka in thousand)

Year	*Profit/Loss*	*Growth Rate*
1995	5307685	-
1996	6036296	13.7
1997	6939984	14.9
1998	7876216	13.4
1999	8936118	13.4
2000	9980883	11.6
Total Average	7512863.666	13.4

Source : Compiled from table 5.43.

Growth Rate of Nationalised Jute mills of Adamjee Zone

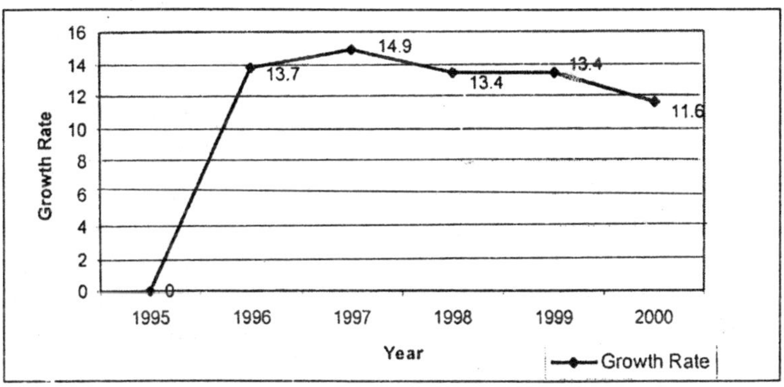

Figure 5.5

In table 5.48 we have made comparison of the rates of profit earned by the nationalised jute mills in different zones of Bangladesh. The table shows that Dhaka zone stands first in earning profit while Khulna zone's position is the worst, *i.e.*, it earns only 8.4 percent profit over the years from 1995 to 2000. Adamjee zone's position is the second just after the Dhaka zone. Adamjee Zone's average growth rate of profit is 13.4 Chittagong zone's average growth rate of profit is 9.9 percent over the years (from 1995-2000). The average rate of growth of profit of the public sector jute mills in Bangladesh as a whole stands for 12.5 percent.

Table 5.48 : Growth Rate of Nationalised Jute Mills Under Different Zones

(Taka in thousand)

Zones	*Growth Rate (Total Average) for Five Years (1995-2000)*
Dhaka	18.5
Khulna	8.4
Chittagong	9.9
Adamjee	13.4
Total	12.5

Source : Compiled from Table 5.43.

However, the data supplied by Bangladesh Economic Review (Bangladesh Arthanaitik Samiksha), 2000, reveals a dismal picture of public sector jute mills. We represent below in table 5.49 the net profit/loss accounts of the nationalised jute mills in Bangladesh for a period of seven years from 1993-94 to 1999-2000.

Table 5.49 is the reality of public sector jute mills in Bangladesh. The table shows that the jute mills in Bangladesh are incurring huge net losses over the period. The net loss situation of Bangladesh nationalized jute mills is shown in terms of growth curve (Fig. 5.6).

Table 5.49 : Net Profit/Loss of Public Sector Jute Mills

(Taka in Crores)

Year	*Net Profit (+)/Loss (-)*
1993-94	64.05 (-)
1994-95	31.43 (-)
1995-96	96.20 (-)
1996-97	251.71 (-)
1997-98	275.71 (-)
1998-99	294.95 (-)
1999-2000	264.45 (-)

Source : Compiled from Bangladesh Economic Review, 2000.

Net Profit/Loss of Public Sector Jute Mills

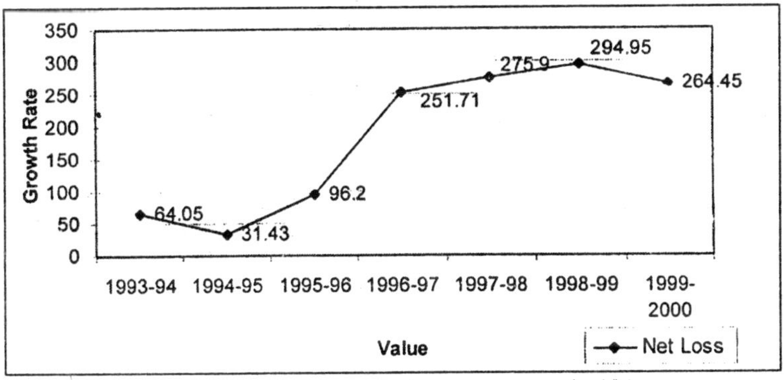

Figure 5.6

Jute Industries Development Strategies

As public sector is losing constantly, private sector has been recognised as an effective means to strengthen the role of it in trade and industry with a view to boost up economic development and augmenting resources for the same. Accordingly, the Government has adopted a comprehensive privatisation policy. In order to materialise the programme the government has laid down detailed procedure. In order to become competitive and viable the privately owned jute industry started operating on a commercial basis in a reformed policy environment. Government has been pursuing the jute sub sector-restructuring programme since July 1993. The Government has already decided its ownership of jute sector investment from 64 percent to 20 percent out of 31 public sector jute mills government has enlisted 23 mills to denationalise. The privatisation board has almost completed evaluation of the asset value of these mills with a view to transfer them to private sector.

In order to achieve the objectives of the fifth Five Year Plan, the following strategies may be pursued for jute manufacturing sub sector during the plan period: (a) Improvement in productivity and capacity utilisation, rehabilitation of the existing jute mills through BMRE programmes, enforcement of industrial discipline, and exploring possibility of participation of labour through profit sharing and participation in management, with better logistic support, privatisation of maximum public sector jute mills and rationalisation of man power therein, improving managerial efficiency through better recruitment and training and improvement in the quality of conventional jute products and reduction in cost of production through eliminating corruption and improving the per loom production, and finally increasing the per acre productivity the target of improvement in productivity and capacity utilization may be achieved, (b) *Diversification of end use of jute.* Jute may be used in diversified way. It could be through the production of pulp and paper, geo-textiles, absorbents, hard boards, handicrafts, mats, decorative wall mats, shopping bags, carpets, blankets. Development and extension of appropriate varieties of jute, kenaf and same fiber to enable round the year supply of these items for establishment and smooth operation of pulp and paper industry will be undertaken. There might be a strategy of the government to attract higher prices of Bangladeshi jute good; she has to improve the quality of her jute products, (c) *Market promotion.* For the extension and development of market of jute goods government has to identify the constraints of the existing marketing system. To overcome those constraints of jute market at home market and abroad proper steps should be taken. After assessing the market demand both at home and abroad, conducting studies on a continuous basis on the jute goods demand with a view to overcoming the demand and supply gap, formulating proper pricing policies, opening, sales centres in Bangladesh missions, abroad, displaying jute goods in the improving countries, participation in the international fairs and contacting foreign buyers to gear up export and formulating effective import and export policies, (d) *Financial and other policy incentives.* This strategy includes rationalisation of monetary and fiscal incentives, viz., reducing interest rate from existing 15 percent to at least 10 percent, exemption to excise duty on electricity, reduction in the prices of oil and gas for generation of electricity for jute mills, and income tax facility may also be given to compete our jute goods in the foreign market, (e) *Strengthening of institutional capability and R & D.* For improvement of quality and yield of traditional jute fibers as well as for diversification of jute products emphasis on R & D should be given. Establishment of a new jute industry development centre under Bangladesh jute research institute (BJRI) should be given highest priority. To evolve new high yielding varieties of jute seeds is a must to make jute and jute goods competitive in the international market, research and development of rami fiber by sharing experience with China will be promoted, (f) Necessary assistance will be sought from the international jute organisation regarding research and development in the areas of jute, jute goods and related products and accessing market as well, (g) *Improvement of environment and natural resources.* Retting of jute is perceived to contribute to water pollution where there is insufficient availability of water, in such areas, canals and small ponds will be dug for retting of jute. Except that new techniques of retting of jute will be explored with minimum adverse environmental effects. Government decision to ban the production of polythene bags to protect environmental disaster, which ensures more and more use of jute. To reduce the acreage under jute and at the same time maintain a higher volume of supply for its diversified and uses, including use of green jute for pulp making, and thus to eare the burden on insufficient forest resources of the country efforts will be made to produce jute through the year through necessary augmentation of soil nutrients, (h) Programmes will be undertaken to provide retrenchment benefits and retraining for self-employment to support affected workers, (i) Planning and implementation of development programmes through local level institution with the introduction of proposed institutional structure at zila, thana and union levels, small and cottage industries planning activities could be included in the functions of such local level institution. During the fifth plan, BSCIC will provide necessary service for industrial development including industry based on jute and will maintain effective liaison with the development programmes of such institutions. The NGOs who have emerged as a new force for development of the rural areas are expected to play an effective role in the skill development of the poor in producing jute goods and varieties of industrial goods, (j) The jute and jute manufacturing sub sector has immense potential for employment generation and poverty alleviation in Bangladesh. Jute production, jute industry and trade, and the related services employ about 10 percent of our labour force and jute industry is the second largest employer in the total industry after textiles, the new and diversified use of jute plants and fibers for pulp and paper, textile yarn blended with cotton and wool, geotextiles, jute reinforced plastic material, luggage's, shopping bags, handicrafts, accessories, etc. will open new avenues of employment. The rural women in addition to their involvement in agricultural sector, will also get jobs in rural areas by utilising finer jute yarn in cottage industries. In the agricultural sector increased yield per unit area will bring down the production cost of jute which will help growers to alleviate their economic hardship by raising their income.

Summary

To sum up the performance of trade unions in public sector jute mills it is clear that the role of trade union is always recognized by management authorities. Trade union also admits the cooperation of management in spite of little differences regarding wages, other benefits and facilities, dispute settlement and workers participation in management. Both the units have good productin level but due to losses workers' satisfaction are yet to be achieved. Recently world jute market situation has stressed developed and better management of the mill which costs better profitability of Rajshahi Jute Mills and Trade Union welcomes the situation with greater endeavor. Withdrawal of polythine shopping bags has simplified the jute goods market. It is expected that through the cooperation of management and trade union jute industries development strategies can easily be implemented as to arrive at national goal and prosperity of our economy, benefiting both the workers community and the jute mills.

REFERENCES

Audited Accounts of Rajshahi Jute Mills, 1996, 1997, 1998, 1999 and 2000.

Bangladesh Jute Mills Corporation (1994), *Quanterly Jute Goods Statistics,* Vol. 82, 3rd Quarter (January March), BJMC, Dhaka.

Bangladesh Jute Mills Corporation (2000), *Annual Jute Goods Statistics, 1999-2000,* Vol. 89, BJMC/Adamjee Com./Motijheel Com. Area.

Government of Bangladesh, (1973), *First Five-Year Plan (1973-78),* Planning Commission, Ministry of Planning.

Government of Bangladesh, (1980), *Second Year Plan (1978-80),* Planning Commission, Ministry of Planning.

Government of Bangladesh, (1985), *Third Five-Year Plan (1985-90),* Planning Commission, Ministry of Planning.

Government of Bangladesh, (1994), *Bangladesh Economic Survey,* 1994.

Government of Bangladesh, (2000), *Bangladesh Arthanaitik Samiksha, 2000,* Ministry of Finance, Government of Bangladesh.

Mahmud, A., (2000), *Bangladesh Economic Scenario,* Hakkani Publishers, Dhaka.

Ministry of Planning, Government of Bangladesh (2002), *Statistical Pocket Book of Bangladesh 2000,* Bangladesh Bureau of Statistics, Government of Bangladesh.

Official Records of Jessore Jute Industries (JJI), Jessore, Bangladesh.

Official Records of Rajshahi Jute Mills (RJM), Rajshahi, Bangladesh.

Veerachamy, R. (2002), *Quantitative Methods for Economists,* New Age International Publishers.

Chapter - 6

PERFORMANCE OF TRADE UNION IN PRIVATE SECTOR JUTE MILLS

Introduction

The most formidable task of restructuring the Bangladesh economy in the post-liberalisation era has been the restructing of public sector industries including the jute mills. In order to accelerate the rate of growth of industries the Government has initiated reform process in the prime sector. The new policy has paved the way for industrialisation under private sector initiative. Thus the role of Government has been shifted from regulatory to supportive in the liberalisation process. The new industrial policy also encourages setting up of new industries under private initiative with the invitation of foreign direct investment. The export-oriented industries have given exemption from tax. This has encouraged the new industries to start production of export-oriented goods (Bangladesh Economic Survey, 1993-94).

Long back since the mid 1970s the Government of Bangladesh realised that state-managed enterprises were not functioning properly. These enterprises had been incurring huge losses due to a number of reasons. To overcome this situation the Government took up the cause of privatising the enterprises. The sole objective was thus to improve the productive efficiency of very input. This was urgently necessary, at that point of time, to attain higher economic growth for the welfare of the toiling masses. Naturally, the process of privatising jute industry began in the late 1970s with the disinvestments of medium-sized jute-spinning mills. These mills were handed over to the private owners. Within a very short span of time thirty five nationalised jute mills were privatised in 1985-86 and by 1992-93 the number rose to thirty nine. The privatised jute mills are as under : Afil Jute Mills, Ajax Jute Mills, Allied Jute Mills, Anowara Jute Mills, A.A. Howlader Jute Mills, A.K. Khan Jute Mills, Bangladesh Fabric Co., Bengal Carpets, Broad Burlap Industries, Chittagong Jute Mfg. Co., Co-operative Jute Mills, Delta Jute Mills, Dhaka Jute Mills, Fauji Chatkal, Gawsia Jute Mills, Jabbar Jute Mills, Janata Jute Mills, Kohinoor Jute Mills, Mashriqui Jute Mills, Mohsen Jute Mills, Moqbular Rahman Jute Mills, National Jute Mills, Newab Askari Jute Mills, Nowapara Jute Mills, Pubali Jute Mills, Quasem Jute Mills, Saleh Carpet Mills, Sattar Jute Mills. S.K.M. Jute Mills, Sonali Aush Industries, Sonali Jute Mills, Star Alkaid Jute Mills, Sultana Jute Mills, Taj Jute Backing Co. Victory Jute Products & W. Rahman Jute Mills (Jute Goods Statistics, 1993-94).

By March, 2003, forty five composite jute mills have been registered under the Bangladesh Jute Mills Association (BJMA), a private Jute Mills Association. The number of Government-managed jute mills is getting reduced and it is learned that ten more nationalised jute mills are going to be privatised very soon (Rashid, 2003).

The present chapter deals with the detailed analysis of two private sector jute mills namely Ajax Jute Mills and Mohsen Jute Mills. We have also presented here the production, growth rate of production, export of jute goods and growth, man-power position of privatised jute mills. We have conducted opinion survey with 130 workers, employees, staff and executives/ managers of two jute mills. We have also gathered opinions from trade union leaders with different political affiliations.

We have again made a comparative study of production, export, man-power position, wage etc. between nationalised and privatised jute mills.

Short History of Two Privatised Jute Mills

This section gives a short history of two private jute mills namely Ajax Jute Mills and Mohsen Jute Mills. The Ajax Jute Mills Ltd. was established at Mirerdanga, Daulatpur, Khulna in 1964. It started production since 1965. It is a medium sized enterprise having 1320 workers of which 1080 are permanent and 240 are temporary workers. The total number of employees and officers is 190 and 59 respectively. It was originally a privately-managed enterprise where 51 percent share were owned by the individuals and 49 percent was the Government share. After liberation in 1971, 80 percent of the private enterprises were nationalised including the Ajax Jute Mills. Till nationalisation it was a profit-making concern. But after nationalisation the industry started experiencing losses. Due to change in Government policy and subsidies offered by the Government the financial position of the enterprise improved substantially and it gradually started making profit. As a policy of the Government this enterprise was denationalised in 1982. After denationalisation the position of this industry again deteriorated. Due to mounting loss the workers are not being provided wage and other benefits properly and timely. The employment absorption capacity of this industry has been decreased. The detail analysis of the jute mills will be given in the following sections just after the overview of the other private jute mills namely the Mohsen Jute Mills.

Other important private jute mills namely Mohsen Jute Mills was established in 1969 which started its production activity since 1970. It was set up at Shiromoni in Khulna with its head office at Nasim Chamber, 21, Dil-Khusa, Dhaka as a private jute industry. The enterprise started its production with sixty five broad looms equivalent to 250 narrow looms. The nationalisation of this industry was made in 1972 which again was denationalised in 1982. The Mohsen Jute Mills is relatively smaller than the Ajax Jute Mills in terms of production capacity, export of jute goods, employment absorption etc.

A Comparison of Nationalised and Private Jute Mills

In this section we present a comparative statistics of production and export of jute goods of nationalised and privately managed jute mills of Bangladesh. We have again shown the man-power position as well as wage differentials of these mills.

The production and growth rate of production of both public sector and private sector jute mills in Bangladesh have been shown in terms of tables 6.1 and 6.2 below.

Table 6.1 : Production of Jute Goods Under Nationalised Mills

(In 000 M.T.)

Year	*Quantity Produced*	*Growth Rate*
1986-87	334	-
1987-88	326	- 2.39
1998-89	317	- 2.76
1989-90	332	+ 4.73
1990-91	261	- 21.38
1991-92	244	- 6.51
1992-93	283	+ 15.98
1993-94	271	- 4.24

Source : Compiled from Bangladesh Bureau of Statistics.

Table 6.1 is portrayed in diagram 6.1 below.

Table 6.2 : Production of Private Jute Mills

Year	*Production (in M.T.)*	*Growth Rate*
1990-91	173438	-
1991-92	172280	- 0.66
1992-93	162865	- 5.46
1993-94	150657	- 7.49

Source : Compiled from Bangladesh Bureau of Statistics.

Table 6.2 is portrayed in diagram 6.2.

Graph Showing Production of Jute Goods Under Nationalised Mills

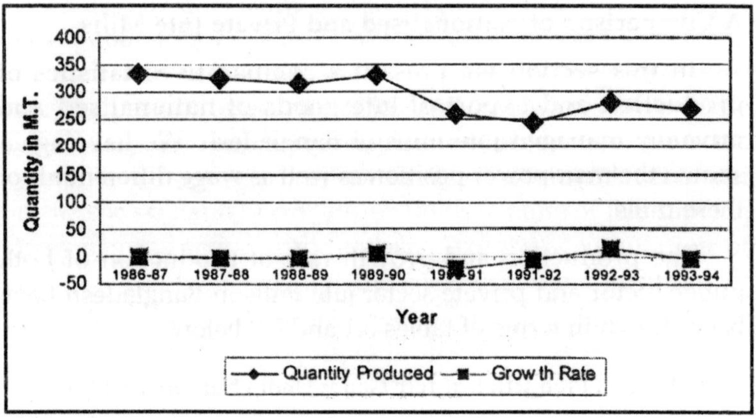

Figure 6.1

Graph Showing Production of Private Jute Mills

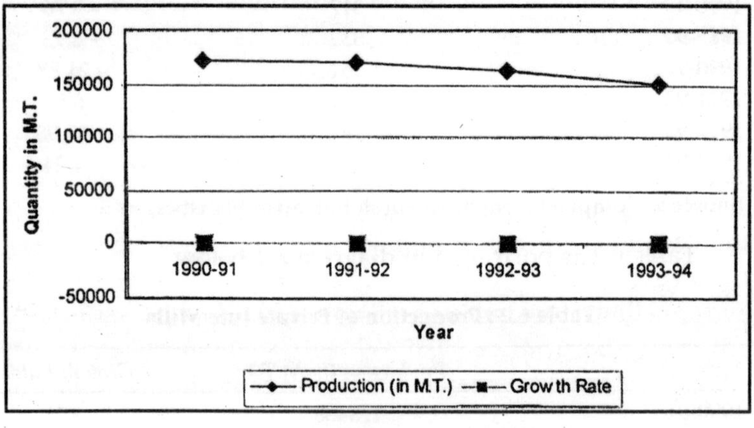

Figure 6.2

The export of private Jute Mills is represented in table 6.3. It is also shown in terms of figure 6.3.

Table 6.3 : Export of Private Jute Mills

Year	*Export (in M.T.)*	*Growth Rate*
1990-91	146740	0.00
1991-92	129116	- 12.01
1992-93	115872	- 10.25
1993-94	118275	+ 2.07

Source : Compiled from Bangladesh Bureau of Statistics.

Table 6.3 is represented in figure 6.3

Graph Showing Export of Private Jute Mills

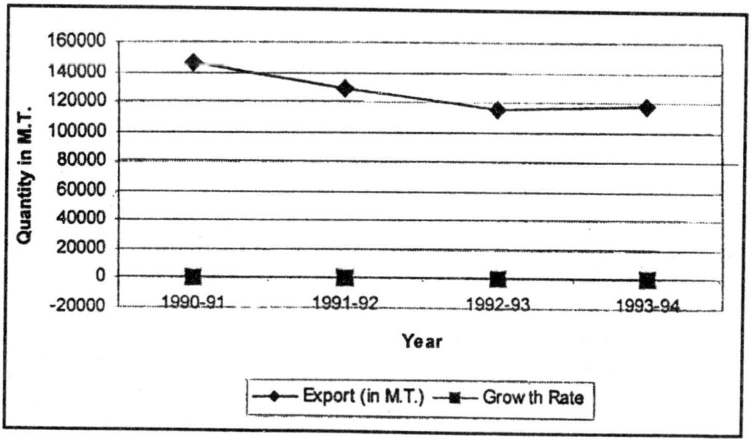

Figure 6.3

A comparative picture of production of nationalised and private jute mills is shown in table 6.4 as well as in terms of bar diagram 6.4 below.

Growth rate of production of both public and private jute mills is shown in table 6.5. This again is represented in terms of growth curve (Figure 6.5).

Nationalised jute mills' contributions were 66.1, 62.43, 64.13 and 69.27 percents in 1996-97, 1997-98, 1998-99 and 1999-2000

Table 6.4 : Production of Nationalized and Private Jute Mills : A Comparison

(In 000 M.T.)

Organization	*Year-wise Production*			
	1996-97	*1997-98*	*1998-99*	*1999-2000*
Nationalised Jute Mills	271.53 (66.1)	255.29 (62.43)	236.19 (64.13)	234.9 (69.27)
Private Jute Mills	139.22 (33.9)	153.59 (37.57)	132.06 (35.87)	104.2 (30.73)
Total	410.75	408.88	368.25	339.1

Source : Annual Jute Goods Statistics, 1999-2000, Vol. 89, Dhaka, Bangladesh.
Figures within the parentheses indicate percentage to total.

Production of Nationalised and Private Jute Mills : A Comparison

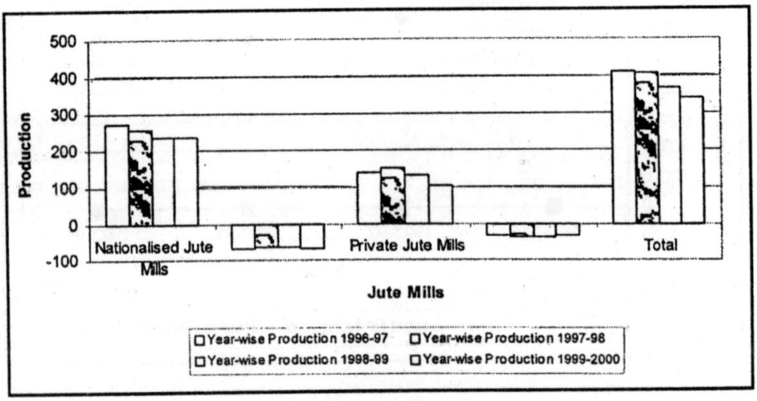

Figure 6.4

Table 6.5 : Growth Rate of Production between Nationalised and Private Jute Mills

Year	*Production Growth Rate*	
	Nationalized Jute Mills	*Private Jute Mills*
1996-1997	—	—
1997-1998	– 5.98	+ 10.32
1998-1999	– 7.48	– 14.01
1999-2000	– 0.54	– 21.09

Source : Calculated from table 6.4.

Growth Rate of Production between Nationalised and Private Jute Mills

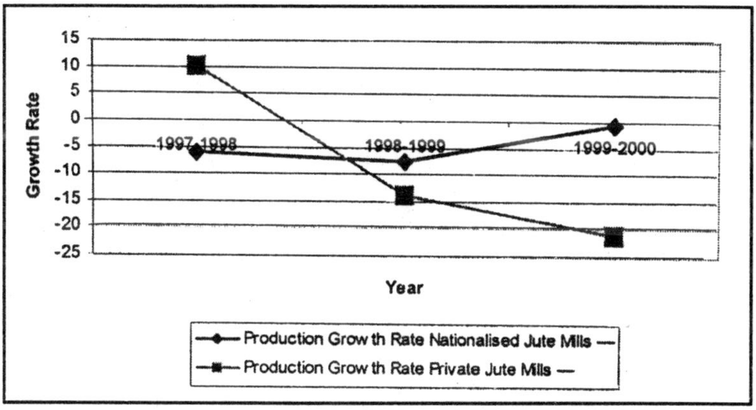

Figure 6.5

respectively whereas the contributions made by the private sector jute mills were 33.9, 37.57, 35.87 and 30.73 percents respectively during the above mentioned periods of time. We receive a dismal picture of growth rate of production of both the nationalised and private sector jute mills shown in table 6.2. In 1997-1998 the rate of growth of production was - 5.98 per cent in nationalised jut mills. It rose to - 7.48 percent in 1998 - 1999 and in 1999 - 200 the nationalised jute mills position was not good enough when they experienced a negative rate of growth of production (-0.54 percent). The position on the private sector jute mills was equally bad. The only exceptional year was 1997-1998 when they experienced a positive rate of growth of production (+10.32 percent). In 1998-99 the private jute mills incurred a negative rate of growth of production of - 14.01 percent. This negative trend continued also in 1999-2000 when this sector experienced a huge negative growth rate of production (-21.09 percent). The growth rates of both nationalised and private jute mills are fitted in figure 6.5.

A comparative picture of export of jute goods and the rate of growth of export of nationalised and private jute mills is shown in tables 6.6 and 6.7.

Table 6.6 : Export of Jute Goods Between Nationalised and Private Jute Mills

(In 000 M.T.)

Organization	*Year-wise Export*			
Nationalised Jute Mills	192.85 (67.92)	160.27 (67.77)	223.84 (73.92)	201.6 (74.91)
Private Jute Mills	91.06 (32.07)	76.20 (32.22)	78.97 (26.07)	67.5 (25.08)
Total	283.91	236.47	302.81	269.1

Source : Compiled from Annual Jute Goods Statistics, 1999-2000, Dhaka, Bangladesh.

Figures within the parentheses indicate percentage to total.

Export of Jute Goods Between Nationalised and Private Jute Mills

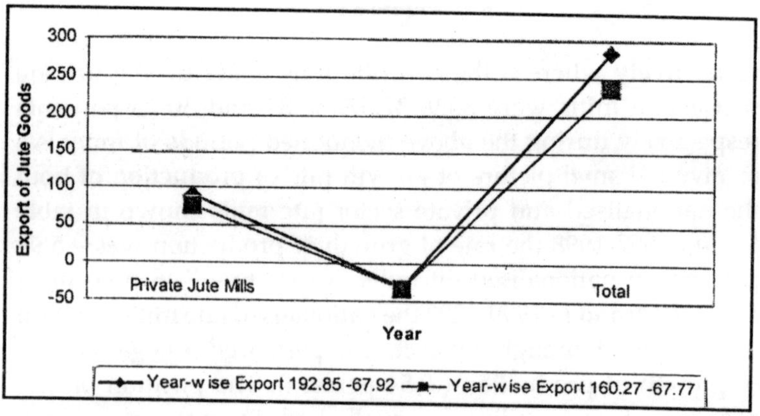

Figure 6.6

Table 6.7 : Growth Rate of Export Between Nationalized and Private Jute Mills

Year	*Export Growth Rate*	
	Nationalized Jute Mills	*Private Jute Mills*
1996-1997	–	–
1997-1998	– 16.89	– 16.31
1998-1999	+ 39.66	+ 3.63
1999-2000	– 9.93	– 14.52

Source : Calculated from table 6.3.

Growth Rate of Export between Nationalised and Private Jute Mills

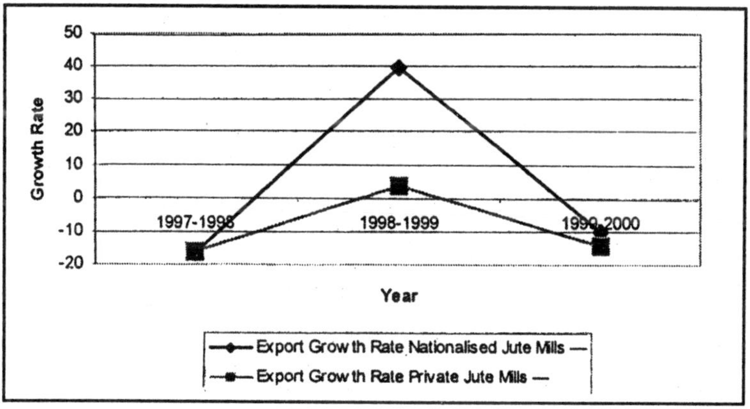

Figure 6.7

Export of jute goods both in nationalised and private jute mills during a period of four years i.e., during 1996-97, 1997-98, 1998-99 and 1999-2000 has been shown in table 6.7 above. This is also shown in diagram 6.6.

The table 6.4 shows that there has been an upward trend of export of nationalised jute mills. The only exceptional year was 1997-1998 when total export of jute goods has fallen from 192.85 thousand metric tons in 1996-97 to 160.27 thousand metric tons in 1997-1998. But the export of private jute mills of Bangladesh has been reduced in terms of total exports as well as in percentage terms. The export scenario of jute mills (both nationalised and private jute mills) has been shown in terms of bar diagrams. The growth rate of expert of both nationalised and private jute mills shown in table 6.4 is negative Annual Jute Goods Statistics showed that in 1997-1998 the rate of growth of export in nationalised and private jute mills were – 16.89 and–16.31 percents respectively. The position improved in 1998-1999 when both the nationalised and private jute mills gained positive growth of export. Nationalised jute mills export recorded 39.66 percent increase in 1998-99 over the previous year. In 1999-2000 both nationalised and private sector mills experienced a huge negative (-9.93 and -14.52 percents respectively) rate of growth of export.

The man-power position of both public and private jute mills of Bangladesh is shown in table 6.5.

Above table shows that nationalised or public sector jute mills are the largest employment providers in the jute industry. This sector provides on an average 59.36 percent employment opportunities to total opportunities available. Total manpower has been divided into two distinct parts, namely, employees and workers. Employees include officers, clerical staff and non-clerical staff. Workers have two separate categories i.e., registered and attendance categories comprising permanent, adhoc, badli and casual workers. Employees provide 59.36 and 40.64 percents respectively to public and private jute mills out of total employment available in this sector. Registered attendance category workers' absorption are 62.62 and 37.38 percents respectively in public enterprises and in cases of attendance category workers these are-74.29 and 25.71 percents respectively. Thus we see that public jute mills in Bangladesh still provide the better employment opportunities than the private jute mills in terms of number of workers employed, wages and other benefits (including social security) provided to the employees and the workers. Opinion surveys also reveal this truth as shown in the following section.

Wages provided to both nationalised and private jute mills workers greatly vary. The nationalised jute workers still receive greater wage and other finge benefits relative to private jute mills. The wage differentials between these jute mills are represented in table 6.6.

The above table 6.9 shows that there is vast wage differentials between the nationalised and the private jute mills in Bangladesh. The pay structure (basic) of nationalised jute mills are more than 2.5 times larger than the private jute mills. Two revisions of wages were made since 1985 for the nationalised jute mills workers, one in 1991 and the other in

Table 6.8 : Man-power Position of Nationalised and Private Jute Mills as on 30.06.2000

Category of Mills	*Employees*				*Workers*							
					Registered				*Attendance*			
	Officers	*Clerical Staff*	*Non-Clerical Staff*	*Total*	*Per-manent*	*Adhoc*	*Badli*	*Total*	*Per-manent*	*Badli*	*Casual*	*Total*
Public	2475	3889	3779	10143	71116	1119	27834	100069	52237	17237	331	699.35
	(56.45)	(61.97)	(58.79)	(59.36)	(67.63)	(100)	(52.0)	(62.62)	(80.68)	(64.17)	(14.18)	(74.29)
Private	1909	2886	2648	6943	34026	-	25688	59714	12502	9696	2003	24201
	(43.55)	(39.03)	(41.21)	(40.64)	(32.37)	-	(48.0)	(37.38)	(19.32)	(35.83)	85.82)	(25.71)
Total	4384	6275	6427	17086	105142	1119	53522	1597	64739	27063	2334	94134
	(100)	(100)	(100)	(100)	(100)	(100)	(100)	(100)	(100)	(100)	(100)	(100)

Source : Compiled from Data Received from Bangladesh Economic Review, 2000, Govt. of Bangladesh.
Figures within the parentheses indicate percentages to total.

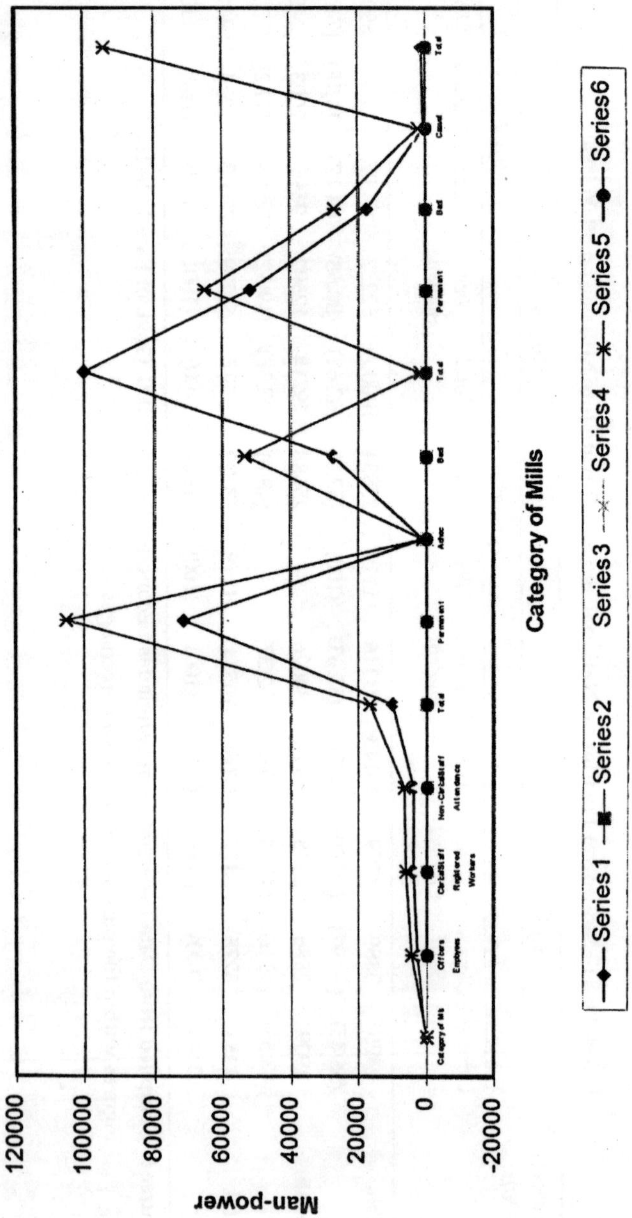
Man-power
120000
100000
80000
60000
40000
20000
0
-20000
Category of Mills
Series1
Series2
Series3
Series4
Series5
Series6

Figure 6.8

Table 6.9 : Wage Differentials between Nationalised and Private Jute Mills

Post	*Basic Pay of Nationalized Jute Mills (In Taka)*	*Basic Pay of Private Jute Mills (In Taka)*
Helper	1550-2270	560-860
Cleaner	1550-2270	560-860
S/Feeder	1600-2410	590-920
T/Feeder	1600-2410	590-920
B/Feeder	1600-2410	590-920
F/Feeder	1600-2410	590-920
Reliever	1600-2410	590-920
B/Shifter	1600-2410	590-920
D.S.O.	1600-2410	590-920
B.M.O.	1650-2550	620-980
Sli : Feeder	1650-2550	620-980
Spinner	1800-2970	730-1180
Reliever Grade	1800-2970	730-1180

Source : Complied from the Record Supplied by the Office of Mohsen Jute Mills, 1997.

1997. But no revision of wages so far was being made for the private jute mills workers. Ironically workers in private jute mills are still drawing a basic pay (along with other allowances) as it was prevailing during 1985 for the nationalized jute mills workers. The socio-economic status of the private jute mills workers are worst than ever. The economic conditions of the nationalised jute mills too are bad but little better than the private jute mills workers. The casualisation of work force in the private jute mills has been increasing fastly especially since the days of privatisation of jute mills. Naturally the role of collective bargaining agents has become almost inactive in settling wage and other benefits.

Like many other developing nations the labour movement in Bangladesh has been supported by multiple trade unions. Each union is again linked to a political party. In fact the pluralism of trade unions has assumed a complex shape. A group of economists and social scientists view that globalisation has benefited the labour class especially engaged in industrial production activities. But one powerful logic that we can put is that the adoption of improved technology and innovation in industries (including the jute industry) reduces labour employment thus reducing employment opportunity in this sector. This extends the quantum of unemployment basket. Under the circumstances labour power becomes cheaper and each worker is given a wage lower than the minimum subsistence wage. This phenomenon is prevalent both in nationalised and private jute mills of Bangladesh. The wage rate prevailing for the jute mills workers in Bangladesh excludes the value of house accommodation, supply of light, water, medical facilities or other amenities. It also excludes the contribution made by the employer to pension or provident fund, traveling allowance or concession or other special expenses entailed by the employment and any gratuity payable on discharge.

An Analysis of Our Field Survey with the Private Jute Mills

This section represents the socio-economic background of trade union members of two private jute mills. We have conducted an opinion survey with the workers, employees, trade union leaders and the management. At the Ajax Jute Mills we have gathered informations from the eighty workers and at the Mohsen Jute Mills we have conducted our survey with fifty jute mills workers. All types of socio-economic data we have collected have been presented in tables 6.7, 6.8, 6.9, 6.10, 6.11, 6.12, 6.13, 6.14, 6.15, 6.16, 6.17, 6.18, 6.19, 6.20, 6.21, 6.22, 6.23, 6.24, 6.25, 6.26, 6.27, 6.28, 6.29, 6.30 and 6.31 where we have show classification of workers on the basis of age (6.7), classification on the basis of education (6.8), parental occupation (6.9), workers encouragement to join the unions (6.10), role of trade unions (6.11), management's attitude towards workers (6.12), ways of placing demands by the workers (6.13), negotiation process (6.14), volume of disputes (6.15), ways of workers facing disputes (6.16), causes of disputes (6.17), nature of workers cooperation (6.18), nature of demand (6.19), workers reaction (6.20), legal provisions followed in the jute mills (6.21), impact of non-complacence with legal provisions (6.22), ways of increasing productivity (6.23), major problems faced by the mills (6.24), major causes of problems faced by the mills (6.25), solutions prescribed by the workers (6.26), nature of working of collective bargaining association (CBA), causes of conflict between trade unions (6.28), suggestions for removing the conflict (6.29) and workers suggestions to management for developing trade union (6.30).

Classification of Workers on the Basis of Age

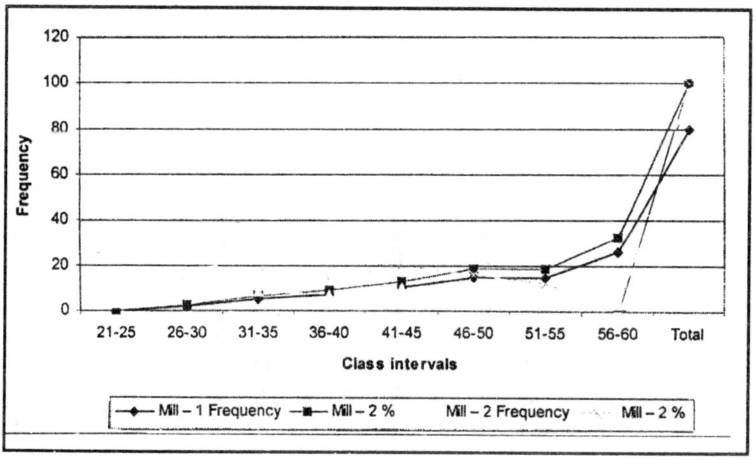

Figure 6.9

Table 6.10 shows that major concentration of workers lies within the age group 56-60. Next concentrations lie in the age groups of 46-50 and 51-5. Data of mill 2 shows that highest concentration lies within the age group 41-45.

Table 6.11 represents that 48.75 percent of the total workers have educational level from Class-VI to Class-X in mill 1 and in mill 2 it is 52.00 percent. No workers were found who received bachelor's degree in mill 1 or mill 2. In mill 1 46.25 percent workers had education upto Class-V and in mill 2 34 percent workers had attained this education level.

Table 6.10 : Classification of Workers on the Basis of Age

Class interval (yes)	*Mill – 1*		*Mill – 2*	
	Frequency	*%*	*Frequency*	*%*
21-25	Nil	Nil	03	6.00
26-30	02	2.50	08	16.00
31-35	05	6.25	08	16.00
36-40	07	8.75	07	14.00
41-45	10	12.50	10	20.00
46-50	15	18.75	08	16.00
51-55	15	18.75	06	12.00
56-60	26	32.50	Nil	Nil
Total	89	100.00	50	100.00

Classification on the Basis of Education

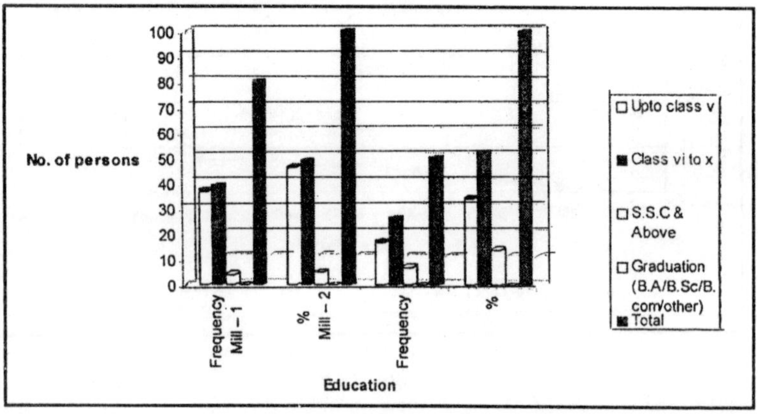

Figure 6.10

Table 6.11 : Classification on the Basis of Education

Level of education	*Mill – 1*		*Mill – 2*	
	Frequency	*%*	*Frequency*	*%*
Upto class v	37	46.25	17	34.00
Class vi to x	39	48.75	26	52.00
S.S.C. & Above	04	5.00	07	14.00
Graduation (B.A./ B.Sc/B.com/other)	Nil	Nil	Nil	Nil
Total	80	100.00	50	100.00

Workers Encouragement to Join the Union

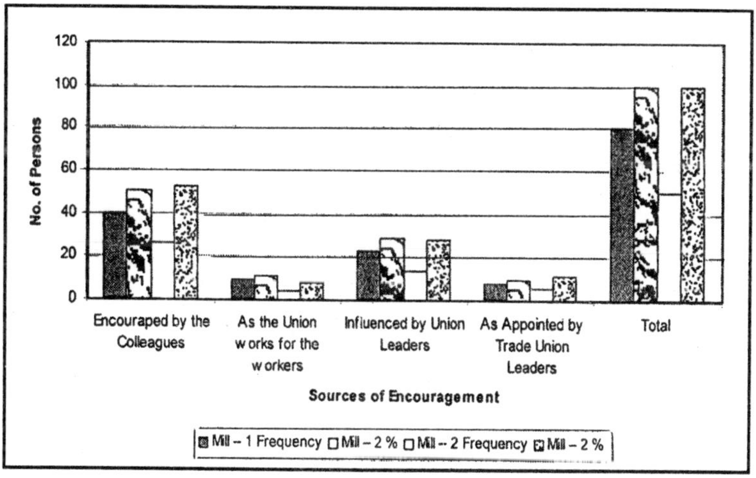

Figure 6.11

Table 6.12 : Parental Occupation

Type of Occupation	*Mill – 1*		*Mill – 2*	
	Frequency	*%*	*Frequency*	*%*
Agriculture	43	53.75	27	54.00
Business	24	30.00	13	26.00
Service	13	16.25	10	20.00
Total	80	100.00	50	100.00

Table 6.12 shows that more than fifty percent of worker's parents have agriculture as the primary occupation on in both mills. The second source of parental occupation is business in mills 1 and 2.

Table 6.13 shows that on an average 50 percent or more of the total workers are encouraged by the colleagues to join the trade union. The opinion survey of mill 1 and mill 2 reveals this fact. Opinion survey of mill 1 shows that 28.75% workers are influenced by the union leaders to join the union and it is 28 percent in mill 2.

Role of Trade Union

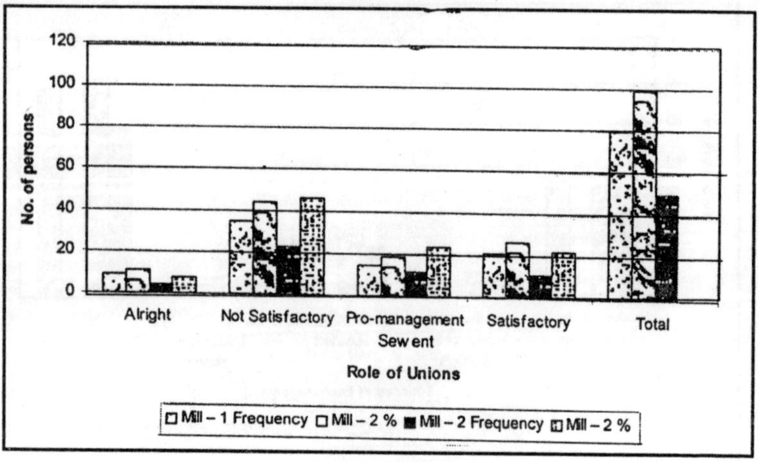

Figure 6.12

Table 6.13 : Workers Encouragement to Join the Union

Source of Encourapement	*Mill – 1*		*Mill – 2*	
	Frequency	*%*	*Frequency*	*%*
Encouraped by the Colleagues	40	50.00	26	52.00
As the Union works for the workers	09	11.25	04	8.00
Influenced by Union Leaders	23	28.75	14	28.00
As Appointed by Trade Union Leaders	08	10.00	06	12.00
Total	80	100.00	50	100.00

As regards the role of trade unions 43.75 percent workers in mill 1 opine that the role of trade union in settling their major issues with the management is not satisfactory. 46% workers in mill 2 support this view. Table 6.11 represents different views of workers regarding the role of trade union.

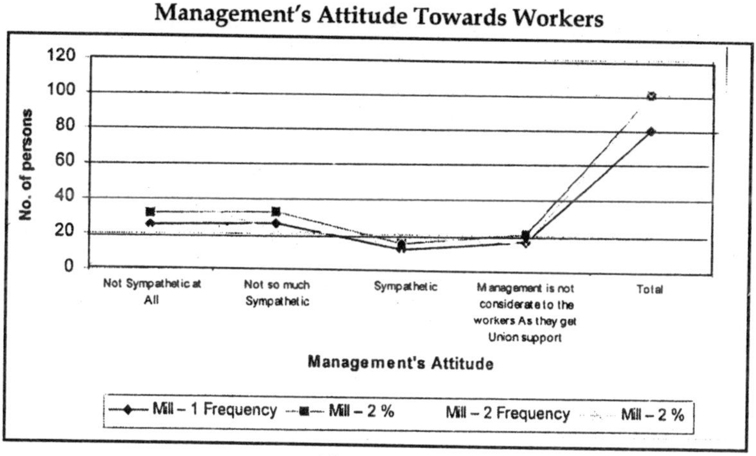

Figure 6.13

Table 6.14 : Role of Trade Unions

Role	*Mill – 1*		*Mill – 2*	
	Frequency	*%*	*Frequency*	*%*
Alright	09	11.25	04	8.00
Not Satisfactory	35	43.75	23	46.00
Pro-management	15	18.75	12	24.00
Sewent Satisfactory	21	26.25	11	22.00
Total	80	100.00	50	

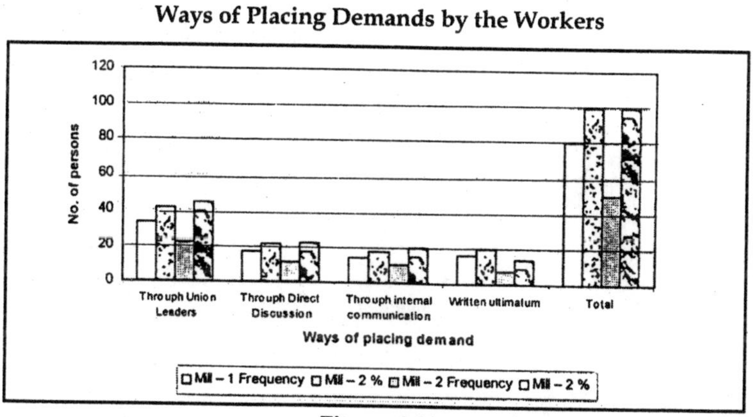

Figure 6.14

Table 6.15 : Management's Attitude Towards Workers

Atitude	*Mill – 1*		*Mill – 2*	
	Frequency	%	*Frequency*	%
Not sympathetic at All	25	31.29	22	44.00
Not so much sympathetic	26	32.50	11	22.00
Sympathetic	12	15.00	09	18.00
Management is not considerate to the workers as they get Union support	17	21.25	08	16.00
Total	80	100.00	50	100.00

Table 6.15 shows that 31.29 percent workers of mill 1 opine that the management is not at all sympathetic to them while 44 percent workers of mill 2 support this view. Only 12 percent and 9 percent workers of mill 1 and mill 2 respectively say that management is sympathetic to their problems.

Volume of Disputes in Last Five Years

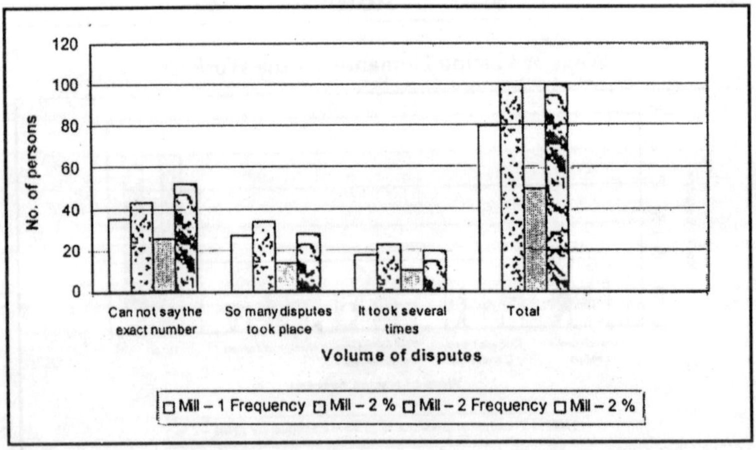

Figure 6.15

Table 6.16 : Ways of Placing Demands by the Workers

Ways	*Mill – 1*		*Mill – 2*	
	Frequency	*%*	*Frequency*	*%*
Through Union Leaders	33	41.25	22	44.00
Through Direct Discussion	17	21.25	11	22.00
Through internal communication	14	17.50	10	20.00
Written ultimatum	16	20.00	07	14.00
Total	80	100.00	50	100.00

Causes of Disputes

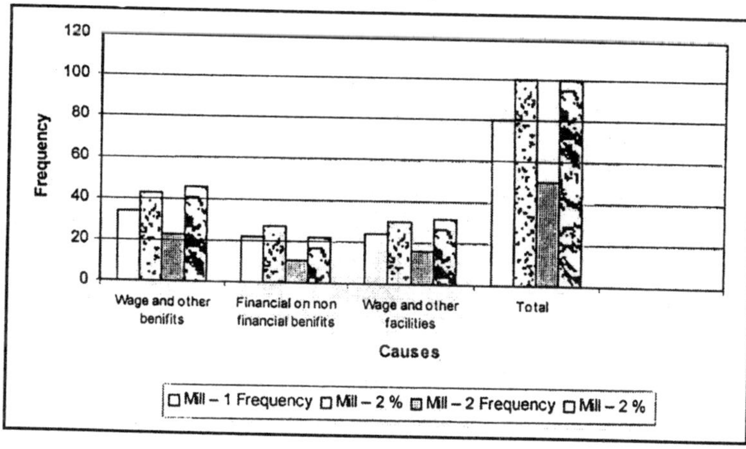

Figure 6.16

Table 6.17 : Negotiation Process

Process of Negotation	*Mill – 1*		*Mill – 2*	
	Frequency	*%*	*Frequency*	*%*
Through C.B.A. Leaders	32	40.00	20	40.00
Through Departmental Heads	15	18.75	08	16.00
Written Ultimatum	17	21.25	10	20.00
Through Strikes	16	20.00	12	24.00
Total	80	100.00	50	100.00

Nature of Demand in Last Five Years

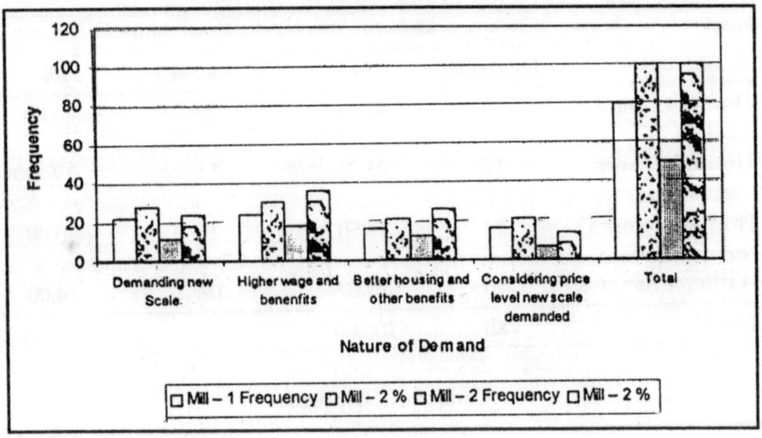

Figure 6.17

Table 6.18 : Volume of Disputes in Last Five Years

Volume of Disputes	*Mill – 1*		*Mill – 2*	
	Frequency	*%*	*Frequency*	*%*
Can not say the exact number	35	43.75	26	52.00
So many disputes took place	27	33.75	14	28.00
It took several times	18	22.50	10	20.00
Total	80	100.00	50	100.00

Volume of disputes took place in last five years is shown in table 6.18 and the ways they face disputes is represented in table 6.19

Table 6.19 : Ways of Workers Facing Disputes

Ways of Facing Disputes	*Mill – 1*		*Mill – 2*	
	Frequency	*%*	*Frequency*	*%*
Through Negotiation Communicable settlement between union leaders of management	18	22.50	20	40.00
Negotiation through C.B.A.	30	37.50	23	46.00
By giving pressure through strikes	32	40.00	07	14.00
Total	80	100.00	50	100.00

Workers Reaction

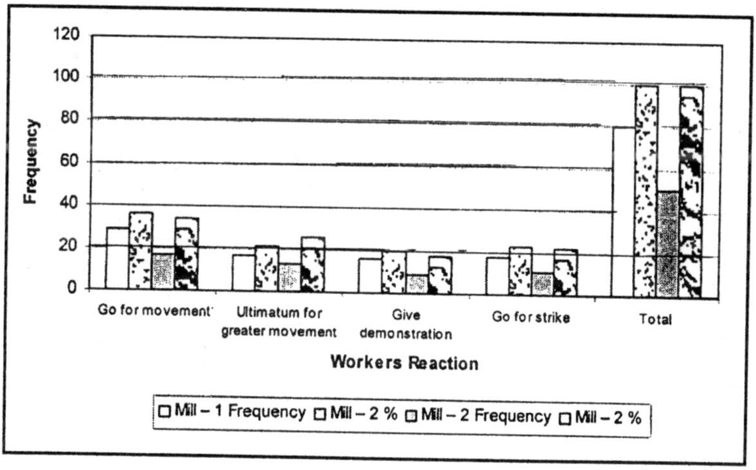

Figure 6.18

Legal Provisions Followed in the Mills

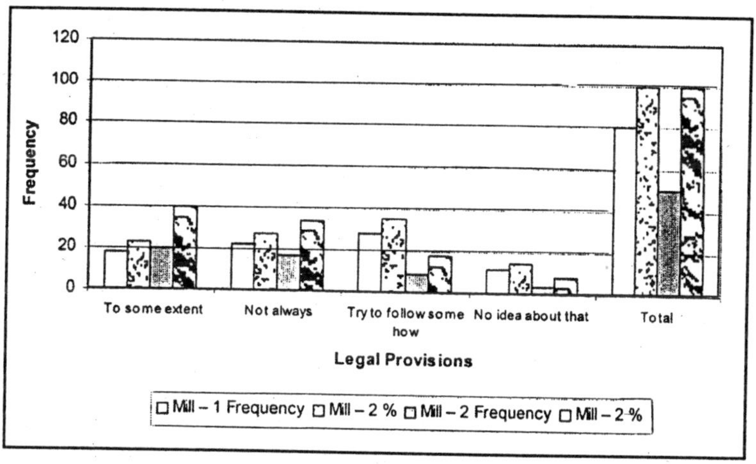

Figure 6.19

Major causes of disputes as opined by mills workers (1 & 2) is shown in table 6.20.

The nature of workers demand is shown in table 6.18 below.

Table 6.20 : Causes of Disputes

Causes of Dispute	*Mill – 1*		*Mill – 2*	
	Frequency	*%*	*Frequency*	*%*
Wage and other benefits	34	42.50	23	46.00
Financial on non financial benefits	22	27.50	11	22.00
Wage and other facilities	24	30.00	16	32.00
Total	80	100.00	50	100.00

Ways of Increasing Productivity

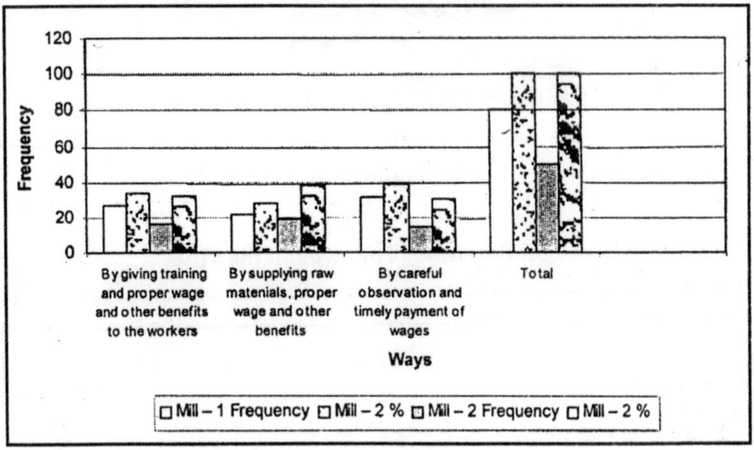

Figure 6.20

Table 6.21 : Nature of Workers Co-operation

Workers' Co-operation	*Mill – 1*		*Mill – 2*	
	Frequency	*%*	*Frequency*	*%*
Workers co-operate at the time of necessities	23	28.75	17	34.00
Co-operate unconditionally	14	17.50	13	26.00
Co-operate without hampering their interest	21	26.25	13	26.00
When management seeks co-operation they go ahead	22	27.50	07	14.00
Total	80	100.00	50	100.00

Major Problem of the Mill

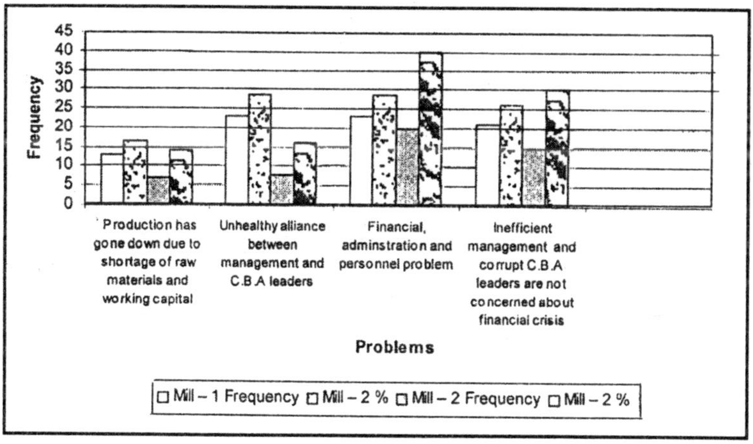

Figure 6.21

Nature of demand placed by the workers of mills 1 and 2 is shown in table 6.22.

Table 6.22 : Nature of Demand in Last Five Years

Demand in last five years	*Mill – 1*		*Mill – 2*	
	Frequency	*%*	*Frequency*	*%*
Demanding new Scale	22	27.50	12	24.00
Higher wage and benefits	24	30.00	18	36.00
Better housing and other benefits	17	21.25	13	26.00
Considering price level new Scale demanded	17	21.25	07	14.00
Total	80	100.00	50	100.00

Table 6.23 : Workers Reaction

Reaction on non fulfillment	*Mill – 1*		*Mill – 2*	
	Frequency	*%*	*Frequency*	*%*
Go for movement	29	36.25	17	34.00
Ultimatum for greater Movement	17	21.25	13	26.00
Give demonstration	16	20.00	09	18.00
Go for strike	18	22.50	11	22.00
Total	80	100.00	50	100.00

Table 6.24 : Legal Provisions Followed in the Mills

Provisions followed in the mills	*Mill – 1*		*Mill – 2*	
	Frequency	*%*	*Frequency*	*%*
To some extent	18	22.50	20	40.00
Not always	22	27.50	17	34.00
Try to follow some how	28	35.00	09	18.00
No idea about that	12	15.00	04	08.00
Total	80	100.00	50	100.00

Table 6.25 : Impact of Non Compliance with Legal Provisions

Impact of non compliance	*Mill – 1*		*Mill – 2*	
	Frequency	*%*	*Frequency*	*%*
Personal or industrial relations will be disturbed	22	27.50	21	42.00
Personal problem will come up	26	32.50	13	26.00
Legal environment will be threatened	15	18.75	08	16.00
Labour problem will emage	17	21.25	08	16.00
Total	80	100.00	50	100.00

The jute mills are facing problems with the level of production and productive efficiency. We have sought opinion from the mill workers. They have suggested the following to improve productivity in the jute mills like (i) providing proper training and adequate wage and other benefits, (ii) supplying raw materials, proper wage and other benefits and (iii) by careful observation and timely payment of wages. In mill 1, 33.75 percent workers say that proper training and wage benefits may help increase productivity. In mill 2, 32 percent workers opine that proper training facility, good wage basket will help increasing productivity.

Private sector jute mills suffer a number of problems faced by the jute mills, among others, are shortage supply of raw materials. This is due to continuous contraction of areas of jute cultivation. The second major problem is the lack of sufficient working capital. Another problem is unhealthy alliance

Total 6.26 : Ways of Increasing Productivity

How to increase productivity	*Mill – 1*		*Mill – 2*	
	Frequency	*%*	*Frequency*	*%*
By giving training and proper wage and other benefits to the workers	27	33.75	16	32.00
By supplying raw materials, Proper wage and other benefits	22	27.50	19	38.00
By careful observation and timely payment of wages	31	38.75	15	30.00
Total	80	100.00	50	100.00

between management and the CBA leaders. The other vital problem is relating to inefficient management.

The major problems of the private jute mills, as told by the workers, are represented in table 6.27.

Table 6.27 : Major Problem of the Mill

Problems of the mill	*Mill – 1*		*Mill – 2*	
	Frequency	*%*	*Frequency*	*%*
Production has gone Down due to shortage of raw materials and working capital	13	16.25	07	14.00
Unhealthy alliance between management and C.B.A. leaders	23	28.75	08	16.00
Financial, administration and personnel problem	23	28.75	20	40.00
Inefficient management and corrupt C.B.A leaders are not concerned about financial crisis	21	26.25	15	30.00
Total	80	Nil	50	100.00

The cause of the problems of mills and the solutions suggested by the workers are represented in tables 6.28 and 6.29 respectively.

Table 6.28 : Major Causes of the Problems of the Mill

Causes of the problems	*Mill – 1*		*Mill – 2*	
	Frequency	*%*	*Frequency*	*%*
Management's consistant negligence to the problems after denationalisation	22	27.50	13	26.00
Main causes are administrative and poor industrial relations and financial	18	22.50	13	26.00
Managements negligence to the official business and poor wage	12	15.00	12	24.00
Main causes are corruption with the help of CBA leaders low wage and other benefits	28	35.00	12	24.00
Total	80	100.00	50	100.00

Table 6.29 : Solution Prescribed by the workers

Causes of the problems	*Mill – 1*		*Mill – 2*	
	Frequency	*%*	*Frequency*	*%*
Management should look after the problems of the mill management and should try to minimise corruption and try to look after the interest of the workers	30	37.50	11	22.00
Management should have to ensure logistic problems of the mill and they have to pay wage timely	08	10.00	03	06.00
Management should look after the problems of the mail and management should try to minimise corruption	35	43.75	13	26.00
Management should try to follow the provision to minimise losses through management audit and should be more careful to make it a profitule organisation	07	8.75	23	46.00
Total	80	100.00	50	100.00

Nature of working of the CBA is shown in table 6.30.

Table 6.30 : Nature of Working of C.B.A

Working of C.B.A.	*Mill – 1*		*Mill – 2*	
	Frequency	*%*	*Frequency*	*%*
Works nicely	06	7.50	09	18.00
Not up to the mark	28	35.00	18	36.00
Not Satisfactory	29	36.25	17	34.00
They serve the purpose of the management	17	21.25	06	12.00
Total	80	100.00	50	100.00

Today the main objectives of the business enterprises are to remain competitive, to be cost-reducing, and to strive for market leadership. Obviously the manager/executives try to control the entire workers community from the very outset. They thus take a strategy to reduce the volume of permanent workers, increase the volume of casual workers, link wage and other benefits with productive efficiency (Janardhan, 2003), increase the number of contract workers, oppose to form trade unions in the plant etc. Multiplicity of trade unions is a major feature in most of the trade unions of the developing countries. Bangladesh is not an exception. In each and every plant there are a number of trade unions, ranging the number from two to ten.

This causes conflicts between the trade unions. Workers of both mills point out that political rivalry, leadership conflict and personality clash are some of major causes of conflicts between different trade unions within the same plant. Some causes of conflict, as pointed out by the mills workers, are shown in table 6.31.

Table 6.32 shows, as suggested by the workers, the remedial measures for reducing conflict within the trade unions.

Table 6.31 : Causes of Conflict Between Trade Unions

Causes of conflict	*Mill – 1*		*Mill – 2*	
	Frequency	*%*	*Frequency*	*%*
Due to political reason	20	25.00	18	36.00
Due to leadership conflict	35	43.75	15	30.00
Due to personality clash	25	31.25	17	34.00
Total	80	100.00	50	100.00

Table 6.32 : Suggestions for Removing the Conflict

Suggestions	*Mill – 1*		*Mill – 2*	
	Frequency	*%*	*Frequency*	*%*
We should try to minimise Politicisation o unions	28	35.00	13	26.00
Early settlement of disputes	24	30.00	16	32.00
Workers participation in management and union activities	16	20.00	11	22.00
Direct negotiation with the	12	15.00	10	20.00
Total	80	100.00	50	100.00

Table 6.33 : Workers Suggestions to Management for Developing Trade Unions

Suggestions to management for developing trade union	*Mill – 1*		*Mill – 2*	
	Frequency	*%*	*Frequency*	*%*
Training for trade union members	20	25.00	10	20.00
Motivation of Trade union members	28	35.00	17	34.00
Close co-operation between management and workers	32	40.00	23	46.00
Total	80	100.00	50	100.00

Our Concluding Observations

As a part of liberalisation process major industries in Bangladesh including the jute industry have been denationalised. The reasons behind such a policy, among others, are to overcome the problem of mounting losses incurred by the jute mills, increase efficiency, reduce average costs and reduce strain relations between the employers and the employees. We have completed twenty years of de-nationalisation process in jute sector in 2002 with a hope of improving the loosing concern. But the conditions of jute mills have deteriorated substantially over this period of time. Of the forty five private jute mills thirty jute mills have been closed. In all, some forty five jute mills which include public jute mills too are now defunct. The industries are closed due to continuous loss. The biggest jute industry in Bangladesh, the Adamje Jute Mills which has already stopped its production for havoc loss and for so many other reasons common for all the jute mills as mentioned earlier. One of our two sample jute mills namely the Ajax Jute Mills also once stopped production. But later on has started production. The golden days of Bangladesh Jute industry (once first position in terms of contribution in the national income) have gone and it seems that those days will not roll back again. Why do we predict this? What are the bases of such a prediction? Is there any internal weaknesses within this sector? Is the sector less competitive in the international market? Is the demand for jute goods declining globally? What roles trade unions, in general and the workers in particular, can play to improve the position of both the public and private jute mills? We have so far tried to answer those questions in chapter -V and in the present chapter. What we can add here is that our jute industries (both public and private) fail to cope with the international competitive conditions; because the jute industry in Bangladesh has not gained any benefit from the advantage of devaluation. Economic theory tells us that the price of a product depends on direct and indirect costs incurred by a firm or an industry while producing that good. These costs are popularly known as average variable costs (AVC) and intrinsic costs of production. This theory does not hold good for Bangladesh jute products. Here the price of jute goods is not fixed on the basis of costs incurred by the mills. The sales price of Bangladesh jute goods is instead decided upon the sales price of competitors in the international market.

The other reason for weak performance of Bangladesh jute mills is that internal market for jute goods are too small. The substitute goods especially the polithin bags have partially replaced the demand for jute goods in Bangladesh. Naturally, home consumption of jute goods is less than fifteen percent of the total jute goods produced in Bangladesh. World statistics of jute goods production and consumption show that jute goods producing countries of the world consume a substantial portion of their own jute goods production. India, for example, consumes almost ninety percent of its total jute goods production. It is necessary to enhance the multiple use of jute goods. This will increase the area of jute production which again will expand the scope of employment in the rural sector.

The liberalisation policy has an impact on industrial relations especially the relation between the management and the workers. Broadly speaking, liberalisation policy throughout the world has three types of impacts. First, globalisation has intensified economic inequality globally. It has adverse impact on employment and wages of unskilled or low-skilled workers and there has been a change in policy orientation on labour relations i.e., relation between the management and the workers. As a result of liberalisation programme in Bangladesh there has been a greater reliance on the market forces of demand and supply, which are expected to play in a greater way the allocative role in the economy. Casualisation of workforce in jute sector has increased, increasing the helplessness among the workers community in this sector. This again has weakened the collective bargaing power of the trade unions. Trade union activities are very week in private jute mills. Naturally, the unions cannot motivate the management to enhance their wages and other benefits. This is due to this fact that last pay revision in the private jute mills were made in 1985. Since then two more revisions were being made in public sector jute mills with modifications of fringe benefits. The private jute mills workers are being deprived of this. They are not in a position to organise movement for better wages and for other benefits. Thus our field survey suggests that the future of private jute mills of Bangladesh is indeed bleak.

REFERENCES

Bangladesh Bureau of Statistics (2000), Statistical Pocket Book, Bangladesh, 2000. Bangladesh Jute Mills Corporation (1993), Quarterly Jute Goods Statistics, 1993-94, Dhaka, Bangladesh.

Bangladesh Jute Mills Corporation (2000), Annual Jute Goods Statistics, 1999-2000, Vol. 89.

Finance Ministry, Govt. of Bangladesh (2000), Bangladesh Economic Review 2000. Govt. of Bangladesh (1985) : Bangladesh Economic Survey, 1984-85.

Govt. of Bangladesh (1994), Bangladesh Economic Survey 1993-94.

Grolier (1999), Encyclopedia No. 31 of Knowledge, Vol. 2, USA.

Hafiz, M.A., (1980), "Trade Union Movement in Bangladesh", Bangladesh.

Janardhan, V., (2003), "Arguing for Industrial Relations : Journey to a Lost World". Economic and Political Weekly August 2-6, Vol. XXXVIII, No. 31.

Kaushal, G., (2001) Economic History of India 1757 to 1966, Kalyani Publishers, New Delhi.

Official Records of Ajax Jute Mills, Mirer Danga, Khulna, Bangladesh.

Official Records of Mohsen Jute Mills, Shiromoni, Khulna, Bangladesh.

Rashid, Harua-Ur (2003), "Bangladesh Jute Industry : Problems and Prospects" Bangladesh Observer, March, 7, Dhaka, Bangladesh.

Seth, N.R., (2001), "Trade Unions in Word and Deed". Indian Journal of Industrial Relations, Vol. 37, No. 2, October.

Srinivasan, T.N., (2002), Eight Lectures on India's Economic Reforms. Oxford India. The Hindu (2002) : India Book of the Year 2002.

Vijay, G., (2003), "Changing Policy Regime and Labour : A Case Study". Economic and Political Weekly, August 2 - 6, Vol. XXXVIII.

Chapter - 7

SUMMARY AND CONCLUDING REMARKS

Summary

Trade union movement was taken birth in response to the changed conditions of the Industrial Revolution. But the movement, at its initial phases, was not successful in implementing the demands of the working class. The movement gained a momentum with the formation of Trade Union Congress and the enactment of Trade Union Act in 1868 in England. In Germany, the workers began to organise after 1848 although their actual growth was noticed after the Second World War. In France workers organised and formed union in the early nineteenth century. The movement became reform-minded since the mid nineteenth century and sought to change the economic and social system rather than simply to bargain with the employers. The American Federation of Labour formed in 1886 pointed out that the basic aim of trade unions was to improve the conditions of the workers through collective bargaining. This idea was, in fact, completely different from that of the socialist idea of trade unions whose objective was, among others, to transform the society.

In the Indian subcontinent the trade union movement began to emerge since the last decade of the nineteenth century with the formation of Bombay Millhands Association in 1890. The early unions were of adhoc nature. Most of them had no regular constitution, no regular membership, no sufficient funds and no constructive programme of actions. A notable development took place with the formation of the Indian National Trade Union Congress (INTUC) in 1947.

To understand the position of the workers of organised sector we have to look at history. Naturally we have made an extensive analysis of the growth and development of trade union movement all over the world in general and a through investigation of the movement in Indian sub-continent, Pakistan (upto 1971) and independent Bangladesh in chapters II, III and IV respectively.

In the discussion of labour movement in the Indian context we have seen how labourers as a class emerged in India and Bengal with the gradual development of factory system. We have also analysed different phases of difficulties faced by the unions both from the employers and the Government. Trade union movement took a new shape after the formation of Pakistan. The movement could not develop unhindered upto 1971 due to division, disunity, personality class, ideological conflict and the control from outsiders.

The growth of the movement in the organised sector has been elaborately discussed with the changing circumstances of major industries including the jute industry in independent Bangladesh. The episodes of nationalisation and again denationalisation of major industries, including the jute industry, have clearly been elaborated. The core of the book has been the representation of two case studies, one with nationalised jute mills and the other with private jute mills of Bangladesh. In the discussion of nationalised jute mills we have tested homogeneity and/or indifference of two jute mills namely the Jessore Industries (JJI) and the Rajshahi Jute Mills (RJM) with the help of a statistical technique known as chi-square distribution (χ^2). We have essentially used this technique to test the significance of the population variation through confidence interval. An interesting picture has been portrayed while analysing the data collected from the two public sector jute mills. The data show that average rate of growth of profit of the public sector jute mills, as a whole, stands at 12.5 percent. But the reality is that the mills are incurring huge net losses over the period. Insecurity of job and low wages have made the workers less efficient and thus less productive. Thus the jute mills are not in a position to secure positive level of net profit. The scenario of private jute mills is more or less same in terms of production, productivity, export, accumulation of profit or loss incurred. But the socio-economic positions of workers are more worse than the public sector jute mills. The trade unions in private jute mills are also less powerful and their collective bargaining power in raising wages and other benefits has been deteriorating continuously.

We have seen that union membership has been reducing rapidly. Bangladesh jute mills are not the only example of this. It is a worldwide phenomenon today. In spite of this trend workers have no alternative but to depend on trade union leaders to place their demands before the management. We have also seen that top-level managements hold anti-union attitude and create pressure on the union activities. Many recent studies also support the view that managers (senior, middle and junior level) have anti-union attitudes. Our opinion survey shows that trade unions have political affiliation and this is used by the leaders to gain the power and the confidence of workers. The study also finds that majority of the workers' want to be affiliated with a union on the plea that union will improve their socio-economic status. We have so far tried to test the hypotheses we have formulated in chapters V and VI.

Concluding Remarks

The Bangladesh jute industry was not making profit during the pre-liberalisation era but that loss was not as mounting as the present jute industry is incurring in the face of globalisation and liberalisation. In the pre-liberalisation era the jute industry was compensated; but the industry at present is getting nothing in the form of bonus voucher. What the sector gets is that it receives a cash incentive of five percent against jute goods export. But severe attacks come from the synthetics products. Moreover paper is a competitor for the jute goods products. Bulk-handling system of goods has again reduced the demand for jute products. The decaying condition of jute industry in Bangladesh has helped increasing poor socio-economic condition of the working class. Due to economic insolvency of the manufactured industry including the jute industry there occurs greater in formalisation of employment. The managements are growing more and more rigid. They are declining to improve the conditions of the workers. They even threaten the workers unions to close down operation if they don't cooperate with the management in regard to settling economic issues. Trade unions are thus not in a safe position. Under the changed circumstances the concept of a secure job comes under severe attack. The message is that there will be a freeze on secure jobs. In such a situation it requires a comprehensive safety net for all the workers engaged both in organised an unorganised sectors. To protect the interests of the jute industry and the interests of the workers engaged in this sector the Bangladesh Government should ban the use of substitute for jute goods products ad encourage the raw jute producers and jute goods producers by providing an enhanced subsidy in addition to the existing subsidy. If this industry is survived from a possible collapse then not only the socio-economic conditions of the workers will be improved but also the millions engaged in agriculture sector in producing raw jutes will be getting protected. This will help stand the national economy on its own resource base. This again will help ushering in a development process for the welfare of all.

SELECT BIBLIOGRAPHY

Abdullah, Faruk, (1983), *Bangladesh Arthonaitik Itihash,* Dhaka University Grantha Sanstha.

Ahmad, Kamruddin, (1978), *Labour Movement in Bangladesh,* Inside Library, Dhaka.

Ahmad, M.S., (1969), *Year Book of Jute,* 1967-68, Art-I, Asian Trade Publications, Dhaka.

Ahmad, M.U., (1997), *Productivity-linked Industrial Wages Agreement in Bangladesh: An Overview,* Report Prepared for ILO (SATT), New Delhi.

Ahmad, Mukhtar, (1978), *Trade Unionism and Labour Dispute in India,* Longmans & Green Co. Ltd., Bombay.

Ahmad, Rakibuddin, (1966), *The Progress of Jute Industry and Trade,* Pakistan Central Jute Committee, Dhaka.

Akkas, Md. Ali & Hossain Ferdous, (1998), Characteristics and Weaknesses of Trade Unionism : Bangladesh Perspective, *Social Science Review,* Vol. XV, No. 2, Department of Sociology, University of Dhaka.

Akkas, Md. Ali (1998), Labour Management Cooperation : The Bangladesh Perspective, *Dhaka University Journal of Business Studies,* Vol. 19 (2).

Alam, Md. Fashiul, (1983), Industrial Disputes Laws in Bangladesh— A Study in their Development Perspectives, *The Cost and Management,* September-October.

Ali, Amin Mohammad and Shams, M. Nisar Ahmad, (2000), Trade Unionism in Bangladesh: A Study of the Impact of the Trade Unionism on Wage, Employment and Work Environment in the

Manufacturing Section, *The Jahangirgar Economic Review*, Vol. 11, No. 1, June.

Ali, Mohammad Wazed, (1975), *Jute in the Agrarian History of Bengal, 1870-1914: A Study in Primary Production*, M. Litt. Thesis, University of Glasgow.

Anstey, Vera, (1952), *The Economic Development of India*, Longmans Green and Co., London.

Audited Accounts of Rajshahi Jute Mills, 1996, 1997, 1998, 1999 and 2000.

Awal, Iftekar-ul (1997), State of Indigenous Industries, *History of Bangladesh*, Vol. 2, Sirajul Islam (ed.), *op. cit.*

Bangladesh Bureau of Statistics, (2000), *Statistical Pocket Book*, Bangladesh, 2000.

Bangladesh Jute Mills Corporation, (1993), *Quarterly Jute Goods Statistics*, 1993-94, Dhaka, Bangladesh.

Bangladesh Jute Mills Corporation (1994), *Quarterly Jute Goods Statistics*, Vol. 82, 3rd Quarter (January-March), BJMC, Dhaka.

Bangladesh Jute Mill Corporation (2000), Annual Jut Goods Statistics 1999-2000, Vol. 89.

Bangladesh Jute Mills Corporation, (2000), *Annual Jute Goods Statistics, 1999-2000*, Vol. 89, BJMC/Adamjee Com./Motijheel Com. Area.

Bangladesh Observer, January 7, 1985.

Bengal District Gazetteers, Calcutta, 1907.

Bengal District Gazetteers, Jalpaiguri, Allahabad, 1911.

Bhattacharjee, Durgadas, (1985), *Industrial Unrest in South Asian Region: A Focus on Theories and Trends.*

Bhowmik, Sarit, (1981), *Class Formation in the Plantation System*, Peoples Publishing House (P) Ltd., New Delhi.

Bhuyan, M. Sayefullah, (1986), Trade Union Movement, in Chakravarty, S.R. and Narain, Virender (eds.) : *Bangladesh*, Vol. II, Domestic Politics, South Asia Studies, Series 13, South Asian Publisher, New Delhi.

Bhuyan, M. Sayefullah, (1991), The Trade Union Politics in Bangladesh: A Study of Textile Workers Union, *Dhaka University Studies*, Part-C, Vol. 12 (1).

Bhuyan, M. Sayefullah, (1991), A Study of Textile Workers' Union, *The Dhaka University Studies*, Part C, Vol. 2 (1).

Bose, Nemai Sadhan, (1969), *The Indian Awakening in Bengal*, Firma K.L. Mukhopadhyay, Calcutta.

Boserup, Easter, (1970), *Women's Role in Economic Development*, George Allen and Unwin Ltd., London.

Broughton, G.M., (1924), *Labour in Indian Industries*, Humphrey University Press, London.

Census of India, 1901, Vol. 1, Part-1, Report.

Census of India, 1911.

Census of India, 1921, Vol. 1, Part-1, Report.

Chakma, P.B., (1991), Industrial Workers Commitment in Bangladesh: A Search for Correlates, *Dhaka University Journal of Business Studies*, Vol. 12 (2).

Chakrabarty, Dipesh, (1996), *Rethinking Working Class in History : Bengal 1890-1940*, Oxford University Press, Bombay.

Chowdhuri Sushil, (1975), *Trade and Commercial Organisation in Bengal (1650-1720)*, Firma K.L. Mukhopadhyay, Calcutta.

Chowdhuri Sushil, (1997), *European Companies and Export Trade, History of Bangladesh*, Vol. 2, ed. Sirajul Islam, Asiatic Society, Bangladesh, Dhaka.

Chowdhury, A. Mannan, (1983), The Settlement of Disputes – A Study of the Jute Industry of Bangladesh, *Journal of Business Administration*, Vol. 14, No. 13.

Chowdhury, S.C., (1987), *Social Cultural and Economic History of India*, Surjeet Publications, Delhi.

Cole, G.D.H. (1962), *A Short History of British Working Class*, Allen Unwin, London.

Cole, G.D.H., (1962), *An Introduction to Trade Unionism*, George Allen & Unwin Ltd., London.

Crouch, Harrold, (1979), *Indian Working Class*, Schin Publications, Ajmer.

D'Costa, Ronald, (1963), *The Role of the Trade Union in Developing Countries (A Study on India, Pakistan and Ceylon)*, Havaux, Nivelles.

Dange, S.A., (1925), *On the Indian Trade Union Movement*, A Communist Part of India Publication, Bombay.

Dutt Romesh, (1902), *Economic History of British India*, London.

Dutt Romesh (1956), *The Economic History of India (Under Early British Rule)*, Routledge and Kegan Paul Ltd., London.

Dutt, Rajni Palm, (1970), *India Today*, Peoples Publishing House, New Delhi.

East Pakistan Labour Journal, 1966, Vol. XIX, No. 4.

East Pakistan Labour Review, Vol. XIX, No. 1, March, 1966.

Eastern Pakistan Labour Journal, December, 1955.

Eastern Pakistan Labour Journal, June 1955.

Eastern Worker, (1968), Vol. VIII, ed. M. Shafi, Bureau of Labour Publication, Karachi, May 3.

Filippo, Edwin B., (1980), *Personnel Management*, McGraw Hill Book Co., New Delhi.

Finance Ministry, Govt. of Bangladesh (2000), Bangladesh Economic Review 2000.

Flenders, Allen, (1962), *The Trade Unions*, Hutchinson & Co. Publishers Ltd., London.

Fryxell, Gerald E. and Gordon, Michael E., (1998), Work Place Justice and Job Satisfaction as Predictors of Satisfaction with Union and Management, *Academy of Management Journal*, Vol. 32, No. 47.

Giri, V.V., (1962), *Labour Problem in Indian Industry*, Asia Publishing House, Bombay.

Gosh, Tushar Kanti, (1978), *Tea gardens of West Bengal*, B.R. Publishing Corporation (Division of D.K. Publishers Distribution (P) Ltd., Delhi).

Government of Bangladesh, (1973), *First Five-Year Plan (1973-78)*, Planning Commission, Ministry of Planning.

Government of Bangladesh, (1980), *Second Year Plan (1978-80)*, Planning Commission, Ministry of Planning.

Government of Bangladesh, (1980), *Statistical Year Book of Bangladesh*.

Government of Bangladesh, (1985), *Third Five-Year Plan (1985-90)*, Planning Commission, Ministry of Planning.

Government of Bangladesh, (1990), *Statistical Year Book*.

Government of Bangladesh, (1994), *Bangladesh Economic Survey*, 1994.

Government of Bangladesh, (2000), *Bangladesh Arthanaitik Samiksha, 2000*, Ministry of Finance, Government of Bangladesh.

Government of Bangladesh, *Bangladesh Economic Survey,* 1984-85.

Government of Bangladesh, *Bangladesh Labour Journal,* 1988 and 1989.

Government of Bangladesh, *Bangladesh Labour Journal,* 1999, Part 19.

Government of Pakistan, (1960), *Trade Union (Amendment) Ordinance.*

Government of Pakistan, *Labour Code of Pakistan (Central).*

Government of Pakistan, (1961), *Trade Union (Amendment) Ordinance.*

Government of Pakistan, (1969), *Industrial Relations Ordinance,* 1969, with commentary M. Shafi, Nov. 1969, Bureau of Labour Publications, Karachi.

Govt. of Bangladesh, (1985), Bangladesh Economic Survey, 1984-85.

Govt. of Bangladesh, (1994), Bangladesh Economic Survey 1993-1994.

Grolier, (1999), Encyclopedia No. 31 of Knowledge, Vol. 2, USA.

Hafiz, S.M.A., (1980), *The Bangladesh Observer,* May 3.

Harper, John F. (ed.), (1987), *Trade Unions of the World,* Gale Research Centre, Detroit.

Hussain, Ismail M., Hye, Syed A. and Ali, A.M., (1996), Structural Adjustment Policies and Labour Market in Bangladesh, ed. Mustafa K. Mujeri, *Centre on Integrated Rural Development for Asia and the Pacific,* (CIRDAP) Dhaka.

Hussain, M.M. and Haque, M.E., (1991), Socio-Economic Background and Characteristics of Trade Leaders at the Basic Union Level, *The Islamic University of Studies,* Vol. 11, No. 1.

Hutt, Allen, (1962), *British Trade Unionism,* Lawrence and Wishart Ltd., London.

ILO Report (1938), *Industrial Labour in India,* Geneva.

Industrial Ordinance, 1969, with commentary, M. Shafi, Nov. 1969, Bureau of Labour Publications, D. Zam Publications, Karachi.

Industrial Relations Ordinance, 1975, Section 4.

Janardhan, V., (2003), "Arguing for Industrial Relations : Journey to a Lost World". Economic and Political Weekly August 2-6, Vol. XXXVIII, No. 31.

Jha, Shiva Chandra, (1970), *The Indian Trade Union Movement.*

Jose, A.V., (2000), The Future of the Labour Movement : Some Observations on Developing Countries, *The Indian Journal of Labour Economics,* Vol. 43, No. 1.

Kaisar, Ashraf, (1992), Business in the Name of Trade Union (in Bengali), *Bichitra,* 7th August.

Karmoker, Shyam Sundar, (1992), Factors Influencing Industrial Relations System (IRS): A Framework for Analysis, *Cost and Management,* March-April, 1992.

Karmoker, Shyam Sundar, (1992), The Quality of Working Life (QWL) in the factories of Bangladesh, *Dhaka University Journal of Business Studies,* Vol. 13 (1).

Karnik, V.B., (1960), *Indian Trade Union–A Survey,* Labour Education Service, Bombay.

Karnik, V.B., (1981), *Trade Union Movement and Industrial Relations,* Somaiya Publications Pvt. Ltd., Bombay.

Karnik, V.B., (1982), *Trade Union Movement and Industrial Relations,* Somaiya Publishers Pvt. Ltd., Bombay.

Kaushal, G., (2001), Economic History of India 1757 to 1966, Kalyani Publishers, New Delhi.

Khan, Abbas Ali, (1992), Pattern and Causes of Industrial Disputes in Bangladesh, *Dhaka University Journal of Business Studies,* Vol. 13 (1).

Khan, Abbas Ali, (1993), Approaches to Resolving Industrial Conflict: A Theoretical Analysis, *Journal of Business Administration,* Vol. 19, No. 1 & 2.

Khan, Abdul Awal, (1990), The Trade Union Leadership and Management in Bangladesh—A Study in their Relationship, *Chittagong University Studies Commerce),* Vol. 6, 1990.

Khan, Hamiduddin, (1980), *The Fundamental Right to Freedom of Association in the Indo-Pak-Bangladesh Subcontinent, University of Dhaka,* Bangladesh.

Khan, Md. Abbas Ali, (1986), Government Policies Towards Labour in Bangladesh : A Historical Analysis, *The Dhaka University Studies,* Part-C, 7 (2): 1986.

Khuda-E-Barkat and Barkat, Abdul, (1992), Growth and Structure of Labour Force in Bangladesh, *Dhaka University Journal of Business Studies,* Vol. 13 (1).

Kling, Blair B., (1966), *The Blue Muting: The Indigo Disturbance in Bengal 1859-1862,* Philadelphia University of Pennsylvania Press, USA.

Mahmood, M., (1968), Trade Unionism in the last ten years, *Eastern Workers*, September, 1968, Vol. VIII, No. 9.

Mahmud, A. (2000), *Bangladesh Economic Scenario*, Hakkani Publishers, Dhaka.

Malik, A.M., (1954), *Labour Problems and Policy in Pakistan*, Pakistan Labour Publications, Karachi.

Mathur, A.S. and Mathur, J.S., (1962), *Trade Union Movement in India*, Chaitanya Publishing House, Allahabad.

Millen, Bruce H.(1967), *The Political Role of Labour in Developing Countries*, The Brookings Institution, Washington D.C.

Ministry of Planning, Government of Bangladesh (2002), *Statistical Pocket Book of Bangladesh 2000*, Bangladesh Bureau of Statistics, Government of Bangladesh.

Mishra, Lakshmindar, (2001), Can Globalisation and Labour Right Co-exists? *The Indian Journal of Labour Economics*, Vol. 44, No. 1.

Mondal, Abdul Hye, (1992), Trade Unionism, Wages and Labour Productivity in the Manufacturing Sector of Bangladesh, Research Report No. 133, Bangladesh Institute of Development Studies (BIDS), Dhaka.

Morning News, Dacca, 8th June, 1962.

Mukherjee, R., (1951), *The Indian Working Class*, Hind Kitab Ltd., Bombay.

Mukhtar, Ahmad, (1935), *Trade Unionism and Labour Disputes in India.* Longmans Green & Co. Ltd., Bombay.

Muniruzzaman, A.K.M. and Shamima, Parveen, (1997), Collective Bargaining in Bangladesh, *The Business Review*, Vol. 3, No. 1 & 2.

Nurullah Sheikh Md., (1985), Patterns of Trade Unions in Bangladesh, *The Rajshahi University Studies*, Part A, XIII.

Nurullah, Sheikh Md., (1995), Industrial Relations System in the Public Sector in Bangladesh : A Case Study of the Textile Industry, Rajshahi University Studies, Part-C, Vol. 3.

Official Records of Ajax Jute Mills, Mirer Danga, Khulna, Bangladesh.

Official Records of Jessore Jute Industries (JJI), Jessore, Bangladesh.

Official Records of Mohsen Jute Mills Shiromoni, Khulna, Bangladesh.

Official Records of Rajshahi Jute Mills (RJM), Rajshahi, Bangladesh.

Pakistan Labour Gazette, Vol. XIV, No. 3, July-September, 1966.

Panadikar, S.G., (1933), *Industrial Labour in India,* Longmans Green & Co. Ltd., Bombay.

Punekar, S.D., (1948), *Trade Unionism in India,* New Book Company, Bombay.

Rahman, Anisur, (1968), *Trade Unions in Pakistan : An Evaluation of Their Problems and Development,* Unpublished Dissertation of Degree of Master of Science (Social Science) of the University of Edinburgh.

Rahman, M.A., (1967), Labour Union and Political Organisations, University of London, *Institute of Commonwealth Studies,* January-May, 1967, No. 3.

Rahman, M.A., (1967), Organised Labour and Politics in Pakistan, collected *Seminar Papers on Labour Unions and Political Organisation,* University of London, Institute of Commonwealth Studies, January-March, No. 3.

Rahman, M.A., (1969), Structure of Trade Unions in Pakistan, *The Pakistan Economic Review,* Vol. 1, No. 1, winter.

Rahman, M.A., (1969), Trade Union Management and Leadership in Pakistan, *Eastern Workers,* Vol. IX, No. 6-7, June-July.

Ramanujam, G., (1986), *Indian Labour Movement,* Sterling Publishers Ltd., New Delhi.

Ramaswami, E.A. and Ramaswami Uma, (1981), *Industry and Labour : An Introduction,* Oxford University Press, Delhi.

Rashid, Harua–Ur, (2003), "Bangladesh Jute Industry : Problems and Prospects" *Bangladesh Observer,* March, 7, Dhaka, Bangladesh.

Rayback, Joseph G., (1966), *A History of American Labour,* Collier Mcmillan Ltd., London.

Report of Administration of Bengal, 1929-30, (1931), Government of Bengal, Bengal Secretariat Depot., Calcutta.

Report of Administration of Bengal, 1932-33, (1934), Government of Bengal, Bengal Secretariat Depot., Calcutta.

Report of the Indian Factory Labour Commission, (1908), Vol. 1–Report and Appendices, Simla.

Report on Workers Education in Pakistan, (1962), *ILO Report,* Geneva.

Review of the Trade in India in 1942-43, Office of the Economic Adviser, Government of India, Published by the Manohar Publications, Delhi.